Did I Ever Tell You About the Whale?

Or Measuring Technology Maturity

Did I Ever Tell You About the Whale?

Or Measuring Technology Maturity

by

William L. Nolte

Information Age Publishing, Inc.
Charlotte, North Carolina • www.infoagepub.com

Library of Congress Cataloging-in-Publication Data

Nolte, William L.
  Did I ever tell you about the whale?, or, Measuring technology maturity / by William L. Nolte.
    p. cm.
  Includes bibliographical references and index.
  ISBN 978-1-59311-963-8 (pbk.) -- ISBN 978-1-59311-964-5 (hbk.) 1. Technology assess-
ment. 2. Technological innovations--Management. I. Title. II. Title: Did I ever tell you about
the whale? III. Title: Measuring technology maturity.
  T174.5.N65 2008
  658.5'14--dc   2008029428

Printed in the United States of America

CONTENTS

© 2006 M. Engling

Preface

I first became interested in measuring technology maturity when Greg Peisert (the "Gregory" part of James Gregory Associates, Inc.) challenged me to come up with a spreadsheet that would calculate and display the technology readiness level (TRL) that a given technology had achieved. This was the genesis of the AFRL TRL Calculator. My earliest effort was extremely simplistic, and I quickly realized that more detail was necessary. The companion software contains a copy of the current version of the AFRL TRL Calculator for your use. This software, an Excel spreadsheet, has been cleared through Air Force channels, and is in the public domain. You will find out more about the TRL Calculator in Chapter 3 and in Appendix D. Appendix E tells you how to install the Calculator on your computer.

As I researched material to improve the TRL Calculator, I learned that technology maturity is a far more complicated subject than it at first appears to be. One day, more in jest than in earnest, I told Dick Lane, my boss at that time, that I ought to write a book on technology maturity. To my surprise and dismay, he said, "Go ahead."

Several years have passed since that fateful day, and you now hold the result in your hands. Along the way, I learned that writing a book is a great way to learn about a topic that you think you already know. In organizing the material, I learned a lot. I also found many gaps in my understanding where I needed to do more research. And I discovered that I was not alone in my quest to understand this subject. I was fortunate to encounter other workers in the field of technology maturity. I was extremely fortunate in establishing a working relationship with some of these. Others I know only through their published work.

As you read this book, may you learn at least some of what I learned by writing it. My greatest hope is that the book might stir interest in the intriguing field of technology maturity and its measurement. As you shall see, there is still a lot of work to be done. Maybe you can be the next big contributor to our understanding of this subject.

STYLE

I tried to use a relaxed conversational style as opposed to a more formal academic style in writing this book. While I do include a few footnotes, in most cases I won't give specific references to published works other than naming the individual responsible for an idea. You can find the full bibliographic details in the Recommended Reading by Chapter list in Appendix B. This convention allowed me to keep the flow of the narrative while still giving you detailed information if you really want it.

I used the whale logo liberally throughout the book. That's mostly because I think it's really cute. I used it as a place holder to fill blank spaces. I think that the whale picture is much nicer than, "This page intentionally left blank," especially since once you've written that, the page isn't blank any more.

It's always difficult in government writing to avoid killing your reader with acronyms. I tried to avoid this tendency, but I know I didn't always succeed. Please excuse any lapses, writing them off as nothing more than a bad case of BEGA.[1] To simplify your life, I tried to introduce each acronym or abbreviation the first time it's used in a chapter, whether you encountered it in an earlier chapter or not. Appendix A is a glossary of all the acronyms and abbreviations used anywhere in the book, except for BEGA. Unless I missed some.

ACKNOWLEDGMENTS

I would first like to thank Michele Engling for designing the whale logo used on the cover and throughout the book. She has graciously allowed me to use her copyrighted work. I think that the whale ties the whole work together and, as I said above, it's cute.

Next, I want to thank Dr. Ruth Geiman for serving as editor and proofreader for this work. She was steadfast in her opinions, and definitely kept me on the straight and narrow path of grammatical correctness. We often fought loudly over word choice, construction, and even spelling. Ruth was invariably correct; I could have saved us both a lot of time by listening to her from the start. Any errors that might have survived her scrutiny are mine alone.

A special "Thank you!" to Jim Bilbro who not only encouraged my interest but also took the time to write both the Foreword and Appendix D. I've been fortunate to work with Jim several times. It's always a pleasure, and he always forces me into the unaccustomed process of thought. I guess that's why I asked him to write the Foreword—revenge. Anyhow, thank you, Jim, for both your work and your encouragement. It's been fun.

I'm not going to try to mention all of the friends and co-workers, from across the services and other government agencies, who helped me understand technology maturity. I would surely leave someone out. I would, however, be remiss if I failed to mention Rob Kruse of Venlogic, LLC, who used to drive me crazy by calling me on the telephone and saying, "Help me understand..." Every time he did that, I was again subjected to the unfamiliar agony of thought. There's also Sid Johnson, whose favorite expression seemed to be, "TRLs are nothing but a red herring." I also received encouragement and knowledge from Mike Sullivan and Matt Lea of the Dayton field office of the Government Accountability Office (GAO).

I'd like to thank my daughter, Tracy, for suggesting the title of the book. Suggesting is too weak a verb. She insisted, nay demanded, that I call it, *Did I Ever Tell You about the Whale?* You'll understand once you have read the Prologue.

I probably also need to formally thank all of my current and past supervisors who supported my work on this book. Here goes a big thank you to Dick Lane, Clare Mikula, Gus Reed, Don Tomlinson, Gary Stanley, and Ken Barker. I'll bet you thought I'd never get it finished, didn't you?

Finally, I wish to thank my wife, Jane, for her support during the writing of this book. Without her help and encouragement, this work would never have come to fruition. I know it hasn't always been easy, but, hey, I'm done now!

NOTE

1.   Bureaucratic experience gone amok.

© 2006 M. Engling

© 2006

Foreword

James W. Bilbro

"Did I ever tell you about the whale?" is the first comprehensive collection of material dealing with the many aspects of technology assessment. I believe it will be an important source of information for anyone involved in the development of complex state-of-the-art systems that, by their very nature, require advancement in technology in order to meet performance requirements. It will also be of vital interest to those involved in research and technology development, providing them with critical information relative to what is required to transition technology from a laboratory environment to an operational one—a tortuous path fraught with difficulty.

The idea that technology maturity, or conversely—technology immaturity, significantly impacts the ability to deliver satisfactory products on time and within cost has existed for many years. As noted in the Chapter on Technology Readiness Levels (TRL's), the idea of defining levels of maturity known as TRL's dates to the 1980's. This was documented in a paper describing the use of NASA's Advanced Research and Technology program as the basis for the U.S. civilian space program (Sadin et al., 1989). This was a significant change in emphasis on the part of NASA, where technology had previously been viewed as merely having a supporting role. This change in role for technology was the result of a revision in the National

Space Policy stating that NASA's technology program "—*shares the mantle of responsibility for shaping the Agency's future—.*" The new emphasis on technology's responsibility was no doubt responsible, at least in part, for a concomitant reassessment of how technologies were developed and infused, with a goal of approaching technology development and infusion in a much more systematic way—one that would increase the likelihood of success. This resulted in the codification within NASA of seven levels of technology maturity (later changed to nine) that described the evolution in technology maturity from initial concept to validation in space.

Although TRL definitions have existed for over 18 years, only within the last four or five years have systematic assessment processes been developed, and terms clarified and standardized such that meaningful assessments can be accomplished. In fact, the concept of the TRL has been adopted internationally with the use of TRL's in Canada, the United Kingdom, and Japan, among other countries. Currently an international working group is attempting to develop an agreement for a set of international TRL's. There has been an additional proliferation of TRL offshoots, including "Design Readiness Levels," "Material Readiness Levels," "Manufacturing Readiness Levels," "Integration Readiness Levels," "Capability Readiness Levels," *ad infinitum*—a process that can be continued *reductio ad absurdum* (Chapter 6)! That being said, all of these offshoots reflect recognition that we are not doing well in the process of developing and infusing technology, and that there are various aspects to the process that must be dealt with in a more propitious manner.

It is perhaps a reflection of the times that this area is taking on such a level of importance even on an international scale. When times are prosperous, such as for NASA in the days of the Apollo Program, there is no particular emphasis on getting things "right" the first time—one follows the "test-fail-fix process," or implements multiple parallel approaches selecting for use the one that is successful.

After the Apollo program, decreasing budgets began to have impacts on NASA's programs and projects. The decades following Apollo saw significant delays and cost overruns. In the early 1980's Werner Gruel, NASA comptroller, produced a now famous curve illustrating the impact of lack of "up-front" investment. His analysis showed that projects investing less than 5% of total project costs early in the program resulted in significant cost growth and schedule slip. Although interpretation of this curve is often focused on requirements definition, in reality, the fundamental problem is immaturity of the technology necessary to meet the requirements. When the required technology is at a level of immaturity such that development costs and schedules cannot be accurately predicted, the result is overall project cost growth and schedule slip. It was in this environment of restricted resources, cost overruns and schedule slips that a revision of the National Space Policy occurred defining the importance of the role of technology development. From that process emerged the idea of TRL's.

Unfortunately, beyond the expansion from seven to nine levels nothing much happened formally within the Agency. Identification of technologies by TRL was required, but readiness level assignment was typically left to the technology developer. Since there was no underlying guidance, evaluations varied widely. Throughout this time individuals proposed refinements and groups implemented their own processes for evaluation, but when DoD was directed to use NASA's TRL process in 2002, they found that a single chart describing the nine levels was all there was. Subsequently, DoD began to develop a myriad of processes, as did NASA, no doubt assisted by Congressional requirements added to the '06 authorization bills of both NASA and DoD specifying that *the technologies required for the program have been demonstrated in a relevant laboratory or test environment;—*. Not a bad incentive!

This brings us back to the subject of this book. Bill Nolte has done the community a great service in creating a comprehensive treatment of the issue of assessing the multi-dimensional aspects of technology. I say multi-dimensional because, as you will see in the course of reading the book, TRL's are just one aspect of technology assessment—they form the baseline. I first encountered Bill while I was initiating a workshop on technology assessment. Bill's contributions to that workshop were of vital importance and our subsequent interactions have proven invaluable to me in my own work. We at NASA's Marshall Space Flight Center have made use of the TRL calculator in the assessment of the Ares launch vehicle. It was the TRL calculator that gave me the idea for automating my contributions to the assessment process—the Advancement Degree of Difficulty (AD^2), a description of which Bill has included as an appendix in this book as well as in his discussion in Chapter 8.

This book and the processes it describes do not provide all of the answers, but they do frame the questions exceedingly well, providing a framework and set of standards that can be used to greatly reduce the ambiguity of technology assessment that has plagued so many efforts to date. I would like to make a few comments which, I hope, will further emphasize and perhaps expand upon some of the aspects of the book. Chapter one begins with a definition of technology, which at first blush would seem unnecessary. I made the mistake of lecturing on technology for several months before realizing that my definition of technology was not that of my audience. One of the most important aspects of the entire process is—define your terms! I would like to add another observation as we consider the question of maturity covered later in the discussion on dimensions. With immaturity comes uncertainty. When a technology is at an early point of development, what is required to "mature" falls in the realm of "unknown unknowns" and this is the aspect *that oft makes calamity of so long life* for both the technology and the product developer! It is the unknown aspect of working out the "bugs" that distinguishes technology development from engineering development. The "unknown unknowns" are further described in uncertainty-based management, as developed by De Meyer et. al., in their paper *"Managing Project Uncertainty: From Variation to Chaos* in the winter, 2002 issue of the MIT Sloan Management Review. De Meyer and his colleagues present a unique perspective on project management that holds out the potential for resolving many of the problems associated with successful completion of complex, hi-tech projects. According to De Meyer, the fundamental idea behind uncertainty-based management is the existence of four different types of uncertainty each requiring different management approaches:

1. variation (the accumulation of small influences that cannot be controlled but can be accounted for);
2. foreseen uncertainty (that which lies within the experiential base);
3. unforeseen uncertainty (that which lies outside the experiential base), and
4. chaos (when unforeseen uncertainty dominates).

Failure to recognize these different types of uncertainty often results in the application of specific management practices which in the end exacerbate problems rather than resolving them. The originators of this concept propose first determining the "uncertainty profile" of a project and then tailoring the management style, planning and execution accordingly. Following up on this idea, I believe that it should be possible to create an uncertainty profile based on technology assessment: that is, what is required to mature the technology necessary to implement a project

And finally I would like to add a word of encouragement for those involved in using Earned Value Management. Application of the processes described in this book provides a critical resource for determining the metrics and milestones necessary for tracking progress in projects requiring technology development. A systematic method of assessing the maturity and subsequently the development requirements will significantly enhance the probability of a successful outcome. Good reading to you!

Prologue

DID I EVER TELL YOU ABOUT THE WHALE?

When they were small, my two children could try the patience of a saint with the single word, "Why?" Even the usual parental response, "Because I said so!" failed to stop the flood of whys flowing from young mouths. In desperation, I came up with the following parable.

"The blue whale is the largest of all mammals. Its mouth is so big that a row-boat would fit comfortably inside, but its throat is only this big around," I'd say, putting the thumbs and forefingers of both hands together to form a circle. "Do you know why?"

"No."

"Because that's the way it is."

After a few, or many, repetitions, all I had to say to stem a tide of whys was, "Did I ever tell you about the whale?"

The kids would immediately reply, "Never mind."

Speaking of whales, this book is not about whales. You won't find a single whale once you get beyond this prologue. Even the whale story you read here is not factual. I made it up. It illustrated a point I wanted to make at the time. If you are a marine biologist, please don't bother telling me about any inaccuracies you may have found in my description of the blue whale. This is not a book about whales. It is a book about technology maturity.

The whale chart in Figure 1 shows how a technology's usefulness changes with time. It represents the technology life cycle, showing how a technology becomes more useful as it gets older up to a point. Then its usefulness starts to decline until the technology becomes obsolete and dies. My concept of technology maturity is based on this chart.

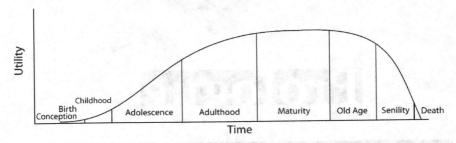

Figure 1.   Stages of Technology Maturity—The Whale Chart.

I introduced the whale into this book because of the shape of this chart. The book will answer your questions about technology maturity and ways to measure it. It will not answer any questions you may have about whales.

The book's repeated references to the whale chart might leave you feeling the same way my kids felt every time they heard, "Did I ever tell you about the whale?"

© 2006 M. Engling

INTRODUCTION TO
TECHNOLOGY MATURITY

PURPOSE

I wrote this book because it fills a need I discovered in my study of technology maturity. There is no standard, generally accepted measure of technology maturity that goes beyond the technology readiness levels (TRLs) originally developed by NASA. This book is a start at filling that void in the technology maturity literature. While the book will not provide the final standard of technology maturity measurement, it is my hope that it will at least serve to spark some interest in this important topic. My specific goals in undertaking this work were the following:

- Define technology maturity in terms of the technology life cycle as demonstrated by the whale chart. I call it the whale chart because of its shape.

Did I Ever Tell You About the Whale? Or Measuring Technology Maturity, pp. 1–18

- Encourage realization of the multi-dimensional nature of technology maturity by developing a proposed set of maturity dimensions.
- Propose possible metrics to be used in measuring the various technology maturity dimensions.
- Promote the use of a single set of TRLs for software as a potential replacement for the several sets of software maturity measures presently available.
- Discuss the problem of accounting for risk during early technology development when you have very limited data. An early risk estimate might be based on a single data point, and that one data point might be little more than your best guess.
- Use the discussion of risk as a jumping off point for talking about current work in measuring the difficulty of further development.

What the Book Is Not

This is not a marketing book. I want to make that clear from the outset. Many authors have written books and papers on technology maturity and product maturity from a marketing perspective. These authors have covered the marketing aspects of this topic in great depth.

What the Book Is

I am not completely ignoring the marketing view. There is some overlap; however, the focus of this book is different. I approach technology maturity and the technology life cycle from a different angle: the viewpoint of the technologist. I'm adopting the perspective of an individual who is developing a technology that may be included on a product at some time.

CHAPTER OVERVIEW

The central focus of this book is technology maturity. This brings two questions to mind, "What is technology maturity?" and "Why is technology maturity important?" I'll answer both of these questions in this chapter.

I'll answer the questions in reverse order. I'll tell you why technology maturity is important to you whether you're a technology or product developer or just a technology user. Then I'll show you what I mean by technology and take a look at the distinction between a technology and a product. This distinction sets the stage for a discussion of technology maturity and product maturity. I'll conclude the chapter by examining two key characteristics of technology maturity: neutrality, the idea that maturity is neither good nor bad, and dimensionality, the concept of looking at technology maturity in several different ways.

WHY IS TECHNOLOGY MATURITY IMPORTANT TO YOU?

Let's divide the world into three classes and see how technology maturity is important to each. You are in at least one of these three classes. Depending on the specific technology and your current job, you might even find yourself in more than one of the three classes. We'll consider each of the three classes in turn:

- the technology developer.
- the product developer.
- the technology user (customer).

Technology Developer

If you're a technology developer, maturing technology is what you do for a living. You need knowledge of the maturity of a new technology so you can estimate the risk and the cost of further development. Knowing where you are on the whale chart tells you how far you've come and also indicates how much more development lies ahead of you before you are finished.

As a technology developer, you'll be finished when the technology has moved far enough to the right along the whale chart. The appropriate level of technology maturity defines the end of the technology development process. When this degree of maturity has been reached, the technology is ready to begin product development. Knowledge of the current state of technology maturity tells you when you are done with your research and development (R&D) program.

Product Developer

If you're the product developer, technology maturity is about risk identification and mitigation. In a 1999 report, the General Accounting Office (GAO) concluded, "The incorporation of advanced technologies before they are mature has been a major source of cost increases, schedule delays, and performance problems on weapon systems." Knowing the current state of technology maturity is vital if you are interested in reducing product development risk. Putting immature technologies into a product increases your development risk in all three of the traditional program management categories:

- cost.
- schedule.
- performance.

Knowledge of technology maturity can also help you avoid manufacturing risk. Knowing how producible a technology is can help ensure that the product can be manufactured using available production processes, equipment, and facilities. An immature technology is more likely to suffer from manufacturing or producibility problems than a more mature technology would experience. Manufacturing problems include low yield, high defect rate, rework, and hand work during production.

At the other end of the technology maturity continuum, you'll face two potential problems by using obsolete technology in a new product:

- performance deficit.
- manufacturing risk.

The performance of a product based on an obsolete technology is going to be less than that of a competing product created with a state-of-the-art technology. Compare the performance of a cassette tape player with that of a compact disk (CD) player. There's no annoying tape hiss during play with the CD technology. It's also easier to get to a particular position in the recording, since you don't have to rewind or fast forward. This performance deficit has resulted in an almost total loss of market share for the now obsolete cassette tape systems. A similar performance deficit could cause a combat disadvantage for the Department of Defense (DoD). Either case, loss of market share or combat disadvantage, presents unacceptable risk to you as the product developer.

If a new product requires obsolete components for its production, the lack of these components can cause serious manufacturing problems. Perhaps you can purchase obsolete components only by paying a premium to get a supplier to build them for you as a custom order. Perhaps the product will have to undergo expensive redesign during manufacturing so you can use available components. Maybe the lack of components supporting the obsolete technology causes you to scrap the entire product.

As a product developer, consideration of risk forces you to know about technology maturity at both ends of the whale chart. You need to avoid using technologies that are either too early or too late in the life cycle.

Technology User

As the customer or end user of a technology or product, you have a different set of priorities. You'll first be concerned about cost and performance, but your long-term concern is supportability. You want to know whether or not you can obtain needed support, spare parts, technology upgrades, or technical advice over the life of the product. A newly acquired item should be serviceable and relatively defect- or bug-free. But after the newness has worn off, you want to continue using it at a reasonable cost for as long as possible. Supportability and sustainability for the long haul become vitally important.

For most products, the major portion of the cost of owning the item comes from operating it and keeping it updated after you have bought and paid for it. You must have some way of keeping the product in an operational state while you own it. If you wish to get more than a one-time use from the item, there must be a reliable and affordable method of making sure the item is available and ready to use every time you need it. Even for single use items, you must have a reasonable assurance that the product will perform properly the one time you need to use it. Supportability or sustainability means that you can be certain that the product will be ready for use when you need it.

What does supportability or sustainability have to do with product maturity? The more mature a product is, the easier it is to keep up. This generalization is true because after a product has been around for a while, the bugs are pretty much worked out of it. A mature technology is more sustainable than an immature technology.

Up to a point, that's true. But many technologies become obsolete. If you use an obsolete technology, you will suffer many of the same problems as the product developer who includes an obsolete technology in a product's design. Spare parts can become scarce, expensive, or unobtainable. The supporting infrastructure might disappear. Technical know-how to maintain the product or provide upgrades and enhancements could vanish. Any or all of these factors could leave you possessing an unsupportable product. Knowing that obsolescence looms in the not-too-distant future can prevent you from investing in a product that will be expensive to support and maintain.

TECHNOLOGY

What is Technology?

Technology is the application of science.[1,2,3] Science and technology are related, but they are not identical. I look at it this way. Science is the search for new knowledge. It is finding knowledge for knowledge's sake. Technology takes the knowledge of science and makes it do something useful for society. When scientific knowledge is put to use, we have a technology. Technology development is the responsibility of the R&D lab. It's in the lab where we take the results of scientific investigation and create something that promises to be useful.

What Kinds of Technology Are There?

When most of us think about technology, the first thing that comes to mind is some piece of hardware. We would all agree that something like the space shuttle, an airplane, a computer, or a DVD player is an example of a technological device. What about nonmaterial technologies? Are there such things? The definitions I quoted in the notes don't rule out nonmaterial technologies.

Software technology is one nonmaterial example. Other nonmaterial technologies include new practices or procedures. These practice-based technologies meet our definitions, since a new practice, process, or procedure is an application of science that allows us to do something in a new way. We could also argue that knowledge itself is a technology. We apply existing knowledge (science) to create and develop new knowledge.

Before we can take advantage of the promised usefulness of technology, we have to take one more step. We have to make something that we can use. The thing we make to put technology to use is called a product.

PRODUCT

William Davidow said, "Great technology is made in the laboratory, but great products are made in marketing." This observation, from his book *Marketing High Technology, an Insider's View*, tells us that a product[4,5] is not the same thing as a technology. While a product can be the result of technology, the emphasis here is on the fact that something has been made. Not all products are technological, but a product must be made.

When you put technology into a product, you productize or commercialize the technology. Productizing a technology makes it useful. This is what your product development staff does. They take a technology developed by R&D and make it useful. If they are good, the result of their work is not only a useful product, but a saleable one as well.

Within DoD, acquisition professionals productize a technology. They are the specialists who take an R&D lab-developed technology and integrate it into a weapon system or install it on an aircraft, space vehicle, ship or ground vehicle. Along with the product development staff referred to above, these specialists are the people whom I call "product developers."

TECHNOLOGY MATURITY

Now that we have an idea of what we mean by technology, let's see what technology maturity might mean. Maturity implies aging, or growth. What do I mean when I say that an inanimate object like a technology matures? You have an intuitive idea of what technology maturity means. You have thought, "I'm not going to buy that thing yet. I'll wait for them to get the bugs out of it." This shows that you have a feeling for what it means for a technology to be immature, not quite ready for prime time. You expect a technology to become more mature, more capable of meeting your needs, after it "grows up."

Growth implies change over time. When I talk about technology growth, I mean that our understanding of the technology improves over time. It's not so much the technology that's changing as it is our knowledge about the technology that's growing. The more we learn about a technology, the better we can apply it to meet our needs. Applying a technology to our needs increases its usefulness, or utility. As we learn to apply a technology, we use it to create the products that fill the store shelves and dealer showrooms around the world.

Our technology could reach the stage where we know almost everything about it. At this level of technology maturity, growth is no longer possible. When a technology reaches this stage, it's not necessarily past its useful life. Many technologies that reach this maturity level continue their existence as part of the common wisdom. Such technologies as the simple machines and fire may have reached this stage. We know most of what there is to know about them, yet they remain useful. There is, however, no apparent competitive advantage to be gained by obtaining increased knowledge of these areas.

This level of maturity rarely comes to pass. That's because most technologies receive a harsher fate. They are overcome by newer technologies that perform the same functions more efficiently or more cheaply. In a word, they become obsolete. Consider how radio tubes became a thing of the past when semi-conductor technology matured. Think about the advances in computer memory devices. Do you remember $5\frac{1}{4}$ inch floppy disk drives? I'll bet your latest computer doesn't have one. Many technologies have been totally replaced by newer ones.

PRODUCT MATURITY

Much of what I said about technology maturity applies here as well. The main difference is that when you look at maturity from the viewpoint of a product, you adopt a marketing view of the world. Product maturity is measured by how long a particular product has been around. When you consider product maturity, you're looking at two different ideas:

- has the product been around long enough so the manufacturer has worked the bugs out of it? In this case, a longer history would be good.
- has the product been around so long that it's in danger of becoming obsolete? In this case, a shorter history would be good.

Like technologies, products can become obsolete. In the product world, obsolescence can come when a product is replaced by a competing product. The competing product may use the same underlying technology, or it could be based on something entirely different. The Pony Express, a service product, became obsolete because of a competing message service, the telegraph. The newer service product exploited a totally new technology.

Products can also become obsolete because of planned obsolescence. When a company introduces a new model of an existing product, the old model is phased out. The annual model changes in the automobile industry are the classic example of planned obsolescence. Last year's model becomes obsolete, not because of new technology, but because the manufacturer plans to produce no more of the previous model.

One measure of product maturity is market share. A new product often has 100% of a small market because competitors have yet to realize that the new market exists. Once the market potential becomes known, competition begins. The market starts to play its economic hand, and the competing products battle for their share of a growing market. As the product market gets saturated, competitors strive to lure their opponents' customers away, since the market is no longer growing. Finally, a few survivors cater to the reduced needs of a declining market.

Some products, such as the light bulb, have avoided the obsolescence problem by becoming part of the social or technological infrastructure. Bulb manufacturers provide a staple of the infrastructure; a commodity that's so necessary that there's little need to advertise. Everyone needs light bulbs, and standardization of the sockets and electricity supply ensure that every manufacturer's product will function in all applications. Products that reach this level of market penetration may avoid the technological death that overtakes most products. The only threat might come from a technology alternative that makes the entire industry obsolete.[6]

MATURITY CHARACTERISTICS

While there are certainly many different characteristics of product and technology maturity, I'm only going to discuss three of them here. The key characteristics I'll cover are neutrality, context dependency, and dimensionality as these terms relate to technology or product maturity.

Neutrality

Richard Turner of the DoD Software Intensive Systems office has stated that the idea of technology or product maturity is a value-neutral concept. In itself maturity is neither good nor bad. A more mature technology is neither better nor worse than one that is less mature. Maturity measures where a product or technology is positioned at a particular time. While you might be able to draw some conclusions about a technology or product based on its maturity or immaturity, the fact that it has reached a certain degree of maturity is just a fact.

Context Dependency

The idea of context dependency is closely tied to the concept of neutrality. It's the situation that dictates whether more maturity is good or bad. It's the circumstances surrounding the maturity level, and the situations under which we're measuring maturity, that define the relative goodness of a particular measurement. The meaning attached to the data comes from the total measurement context.

Your purpose in measuring the maturity of a particular item affects the meaning of the measurement. If you need to reduce the risk of a new product development, you would look for a well-understood, mature technology. If you're trying to see whether the technology is ready to leave the laboratory and be added to a product, you'll need at least a minimum level of maturity. In either case, more maturity would be better. If you're considering whether to buy an existing product or invest in developing a new one, you need to worry about the impending obsolescence of a too mature technology. You have now moved into the "more technology maturity is bad" area.

Dimensionality

What do I mean by the dimensionality of an idea? Technology maturity is a concept, not a material thing. How can I talk about the different dimensions of a thing that does not exist in normal, three-dimensional space? What do I mean when I say, "Technology maturity exists in many different dimensions?"

The dimensions of technology maturity are not the three spatial dimensions we know so well. When I speak of dimensions in the context of technology maturity, I mean different ways of looking at or measuring technology maturity; different viewpoints or perspectives. If you observe technology maturity along only one of its dimensions, you'll see an incomplete and erroneous view of the whole. A full picture of technology maturity requires looking at and measuring maturity from all possible viewpoints, including:

- technology maturity.
- programmatic maturity
- developer maturity.
- customer maturity.

Technologists often ignore the last three viewpoints, maybe because they don't think that administration and marketing are as much fun as technology development.

Technology Maturity Dimensions

I will cover six viewpoints or perspectives for looking at the maturity of the technology itself. My six technology maturity dimensions are:

- Current state of technology development.
- Amount of development work remaining.
- Difficulty of remaining work.
- Predicted supportability of final product.
- Interoperability with existing systems or products.
- Manufacturing and producibility.

You may place other technology dimensions on your personal list or you might combine these six in different ways. My six dimensions include elements containing both historical and predictive perspectives. These two perspectives can change, however. As a technology matures, a particular dimension could move from the predictive category to the historical.

That's why I look at the six dimensions individually instead of combining them into two dimensions: historical and predictive.

Current State of Technology Development. The first technology dimension is always historical. It gives you a snapshot of the current state of the technology, telling how far the development has progressed. It shows what has been accomplished and indicates what is left to be done, but does not give any information about how much effort or risk there is in advancing further in the development process. Knowing the current development state gives you a lot of information; however, it tells you nothing about the difficulty of further progress.

One useful measure of technology development is the TRL developed by NASA. The TRL is a 1 through 9 scale that shows how mature a technology is at a particular time.

Amount of Development Work Remaining. This dimension follows from the previous one. If you know the current state, and you know what's expected, the difference is the amount of work remaining. The work still to be done can tell you how far along you are in the development process.

I'll give you an example to make this idea clear. You have two requirements. You are required to demonstrate 20 watts of output power from your device. You also have to get the weight down to 20 pounds. If your current lab mock-up produces 16 watts and weighs 30 pounds, you are 20% short of your development goal for power. You are also 50% over the weight requirement. If you combine these two values,[7] you'll find that you have approximately 32% of the development effort remaining.

In the software world, one measure of remaining effort is an estimate of the amount of code left to write, given either as source lines of code or function points remaining. A required increase in the scale of the existing version of the software might also be a measure. You may have demonstrated its functionality with ten users, but now need to scale it up so thousands of people working on-line can all use it at the same time.

Do you see the problem with using the amount of work remaining as an indicator of technology maturity? Knowing how much work you have left tells you nothing about how hard it will be to get that last required performance increase.

Difficulty of Remaining Work. If we try to measure the difficulty of the work remaining, we're peering into a predictive dimension. The main drawback here is that any prediction is uncertain. Nevertheless, making the effort often provides you with great insight into a program. NASA's John Mankins created the Research and Development Degree of

Difficulty. Mankins defines this as "A measure of how much difficulty is expected to be encountered in the maturation of a particular technology." Note the word "expected."

Jim Bilbro, also from NASA, has extended Mankins' work into the Advancement Degree of Difficulty (AD^2). He created a matrix that quantifies the predicted difficulty you can expect in trying to advance a technology from its current maturity level to the maturity desired. Bilbro asks questions in four pertinent areas to create his AD^2 assessment: design and analysis, manufacturing, test and evaluation, and operability.

Predicted Supportability of Final Product.   The supportability or sustainability dimension is made up of three sub-dimensions. These are estimates of the reliability, availability, and maintainability the production item will demonstrate after it is fielded. The supportability dimension changes from a predictive measure of maturity to a historical measure as development proceeds. This is because, as you develop the technology, you will be able to demonstrate its achievements in each sub-dimension. Let's take a closer look at these three.

- Reliability is the probability that the item will be fit for use at any given time. Ideally, this statistical probability would be drawn from a sufficiently large amount of data. Early in development you must estimate this probability because you don't have enough data to extrapolate it. As development progresses, you may be able to measure the item's reliability performance in the lab. The reliability demonstrated by the lab mock-up can then be used to predict later fielded performance. If demonstrated reliability fails to meet program requirements, a formal reliability growth program may be needed.

- Availability is the probability that the item will be fit for use when it is needed. This measure adds operational need to the reliability measure discussed above. As the item goes through development, availability can be demonstrated through a formal test and evaluation program.

- Maintainability is the ease with which an item can be repaired after it breaks. Computer modeling and simulation can provide early estimates of maintainability. Later in development, maintenance demonstrations can show that required maintainability goals have been met.

Interoperability With Existing Systems or Products.   You could also attempt to estimate whether the technology or product can operate seamlessly with other items already in use. This is the interoperability dimension. Interoperability measures how compatible systems or components are, especially when they must exchange information or interact with each other.

Unfortunately, there is no universally accepted measure of interoperability. The best you can do when working on a technology that requires interoperability is to carefully design and manage the physical and data exchange interfaces. This includes both internal interfaces within your system or technology, and the external interfaces that connect your technology to the outside world.

Manufacturing and Producibility. Producibility is all about whether or not you can manufacture a product that uses a particular technology. A product or technology that cannot be produced is immature to the point of uselessness. While this immaturity could have something to do with the product itself, it is more often a result of the current state of the manufacturing industry's production capability. Improving manufacturing capability is a method of bringing a new technology to use. Producibility can be measured using the Manufacturing Readiness Level (MRL) scale which is analogous to the TRL discussed above.

All of the dimensions and measures I've discussed up until now deal with the technology or product itself. Now we'll look at three other dimension categories which are part of a complete picture of technology maturity:

- Programmatic
- Technology/Product Developer Maturity
- Customer Maturity

Programmatic Dimensions

The first other dimension category includes items that are programmatic in nature. In any technology development program, the program manager has to be concerned with many details that have nothing to do with the state of the actual technology. Attention to these details can, however, spell the difference between life and death for the program. In this section, we'll look at three programmatic dimensions of maturity.

- Documentation.
- Customer focus.
- Budget.

Documentation. The job's not done until the paperwork is finished. If you as a program manager don't keep up with your documentation, you can't show that the technology is performing correctly. Documentation includes a lot of things. In some cases, it might be a technical report. It could also include equipment records, such as the calibration status of test equipment used in a technical investigation. It may include program management planning documents such as a systems engineering plan or a test plan. It could include the agreement to pass on a technology that has completed laboratory development to the next level of product development, production, or acquisition. Sometimes we include the end user or customer as one of the coordinators on a technology transition planning document.

Customer Focus.   Every successful technology development must have a customer focus. The technology must meet a customer's needs or the customer won't use it and won't buy or invest in it. A customer focus requires you to develop a marketing mentality. First, you'll identify potential customers. It's impossible to negotiate a technology transition agreement with your customer if you don't know who the customer is. Then you'll show your customers that the technology satisfies one or more of their needs or requirements.

You must look at the technology through your customers' eyes to show that it performs some function that meets their desires. One way to do this is to include a customer representative on the technology development team. If customers are involved early, they are likely to become committed to the technology under development. Ideally, they will express their commitment by bringing money to the program, either as an investment in the research and development or as a contract to buy the completed product.

Budget.   Talking about money is a great segue to our final programmatic dimension. Budget management as a measure of technology or product maturity may seem a little strange. True, it has little to do with the state of the technology; however, if you do not take care of budgetary concerns, your program is doomed. Measuring the budget dimension usually means adhering to the financial management system used in your organization. It can be as simple as reporting on expenditures or it could require an activity based costing system complete with earned value measures. In any case, you must take care of the budget to be successful.

Money available and money spent are the measures of choice to the bean-counters of the world. Fiscal measures give managers a feeling for where a technology is in relation to the development plan, and this can be a measure of overall technology or product maturity.

Technology/Product Developer Maturity Dimensions

When considering technology or product maturity, we also need to think about the technology or product developer's maturity. We'll look at three dimensions that could fall under the developer category:

- Capability Maturity
- Process Maturity
- Past Performance

Capability Maturity. The capability to perform measures the developer's maturity. The Software Engineering Institute of Carnegie-Mellon University has developed a formal method of measuring a software developer's capability with their Capability Maturity Model (CMM), a five level evaluation of the developer's ability to repeatedly and consistently deliver quality software products. They have expanded this model into the Capability Maturity Model for Integration (CMMI) to include the integration of software and hardware technologies. CMM and CMMI measure a developer's capability based on an evaluation of the personnel and processes present in the organization.

Process Maturity. Another possible measure of capability is concerned only with the production process. This measure is process maturity. It is especially useful in evaluating the capability of a manufacturing process to produce an acceptable product. Process maturity is applicable to both hardware and software. It is an element of the CMM capability metric.

Process maturity can fit in either of two maturity categories. If you include process maturity in manufacturing readiness, it fits under producibility, one of the technology dimensions. Putting it there removes it from the developer maturity dimension. This shows that the technology maturity concept is still under development. Until accepted standards have been created and disseminated, there will be conflicting interpretations.

Past Performance. We can also estimate a developer's capability by looking at past performance. If a company or individual has a good track record of delivering comparable products or technologies as promised, that record can be evidence of the developer's capability to perform. Of course, the opposite also holds. If someone has shown an inability to deliver as contracted, you should view that developer as a risky supplier of a needed technology or product.

Customer Maturity Dimensions

The other key participant in a technology or product development transaction is the customer or end user. Unless the customer is ready to receive the new technology, the deal is dead before it gets started. How can you measure an individual's or organization's readiness to receive a new technology or product? Take a look at four possible customer maturity dimensions:

- Chronological Age
- Enterprise Maturity
- Infrastructure
- Corporate Culture

Chronological Age. In the individual case, the marketing or sales department is responsible for doing the market research that tells you what the customer needs and wants. Here is where customer maturity comes into play. The marketers consider the age of the prospective customer when they design their sales campaign. Unfortunately, when your customer is an organizational entity, considering customer maturity requires more than knowing the organization's chronological age.

Enterprise Maturity. When your customer is an organization with a bureaucratic soul like the DoD, you'll need to consider the maturity of the entire enterprise you're selling to. Enterprise maturity is not easily measured, especially from outside the organization. In measuring enterprise maturity, you are trying to assess the total organization's technical knowledge and understanding. You will use this assessment to gauge the ability of the organization to assimilate the new technology you're offering. Knowing what technology, software, tools, or processes the customer is currently using tells you a lot about the enterprise maturity. You'll still face a high degree of uncertainty, however, when you extrapolate from this knowledge to an estimate of their overall readiness to adopt a new technology or product.

Infrastructure. One thing that affects a customer's readiness to receive a new technology is the infrastructure that the customer presently possesses. If a new product requires the scrapping of an entire information technology network, for example, that makes it tough to sell that product. It's easier to sell a product that fits into the organization's present system.

Corporate Culture. Corporate or organizational culture is an important consideration in estimating the customer's maturity and readiness to receive new technology. An organization with a very conservative culture would favor keeping the technology that it already knows. An innovative mindset marks an organization that more readily adapts to new technology.

The culture will also affect how the organization goes about introducing a new product. Some start slowly, try it out in a small group, work the bugs out, and then gradually expand the use of the new product. Other cultures may support an all at-once-approach. The enterprise culture can either enhance or inhibit the organization's ability to adopt new technology.

CONCLUSION

You have seen that technology maturity is important to you, whether you are the technology developer, the product developer, or the technology user. In every case, knowledge of the state of technology or product maturity provides you with valuable insight.

As I have been talking about technology maturity, I have been talking around the idea of a technology life cycle. In the next chapter, I'll show you a detailed picture of the technology life cycle. If you understand the technology life cycle, you'll understand technology maturity. Did I ever tell you about the whale?

NOTES

1. The American Heritage Dictionary—(a) The application of science, especially to industrial or commercial objectives. (b) The entire body of methods and materials used to achieve such objectives.
2. The MSN Encarta Dictionary—(1) **"application of tools and methods:** the study, development, and application of devices, machines, and techniques for manufacturing and productive processes, and **(2) method of applying technical knowledge:** a method or methodology that applies technical knowledge or tools."
3. NASA Technology Plan "Technology is defined as the practical application of knowledge to create the capability to do something entirely new or in an entirely new way. This can be contrasted to scientific research, which encompasses the discovery of new knowledge from which new technology is derived, and engineering which uses technology derived from this knowledge to solve specific technical problems."
4. The MSN Encarta Dictionary—(1) "something made or created: something that is made or created by a person, machine, or natural process, especially something that is offered for sale (2) company's goods or services: the goods or services produced by a company."
5. The American Heritage Dictionary—"Something produced by human or mechanical effort or by a natural process."
6. Jim Bilbro of NASA points out that solid state light sources and halogen bulbs are already starting to replace incandescent light bulbs in some

applications, so the light bulb might be an example of an infrastructure technology that's facing obsolescence in the not-so-distant future.

7. Garvey and Cho present one method of rolling up these values in their paper listed in the Recommended Reading List by Chapter, Appendix B.

© 2006 M. Engling

THE TECHNOLOGY LIFE CYCLE

CHAPTER OVERVIEW

In this chapter, I'll tell you about the whale by describing the technology life cycle graphically in "The Whale Chart." First I'll show you a biological system as an analog of a technological system. A biological system (life form) undergoes several more or less distinct stages during the time it exists. Similarly, technologies develop along a predictable path from initial idea through usefulness to obsolescence. It's this path of development and usefulness that I call the technology life cycle. I'll talk about some alternative descriptions of the technology life cycle, updating the Whale Chart as I go. The chapter concludes by discussing some exceptions to the life cycle.

Did I Ever Tell You About the Whale? Or Measuring Technology Maturity, pp. 19–33
Copyright © 2008 by Information Age Publishing

BIOLOGICAL METAPHOR

The biological model I'm using goes through the following stages: conception, birth, childhood, adolescence, adulthood, old age, senility, and death. I am not going to expand on the biological meaning of these stages. Take them as given. I'll put you in the inventor's shoes as I demonstrate how your new technology behaves as it goes through each of these stages of maturity in turn.

Conception

This phase is defined as the time when you first get an idea for a new technology. The stage starts with your first glimmer of an idea. You think, "There must be a better way to…" The technology didn't exist; now it's an idea, a thought, a supposition. The new technology doesn't have any definite use yet. It's all pretty nebulous. You don't know whether the idea is worth anything. You're not sure that the concept will work, and if it does work, you haven't figured out how to use it. In this very early stage, the technology concept is poorly formed.

The conception stage can take a very long time. This is because, in this stage, you are going through the entire incubation of an idea. Much of this phase consists of preparation, as you gather scientific or technical knowledge that may or may not be helpful. The hard work in the conception phase takes place in your mind, as you work out the basics of the problem.

Birth

The birth phase begins like a bolt from the blue with the "Ah, ha!" or "Eureka!" experience that suddenly makes everything seem clear and obvious. Typically the birth stage is shorter than the conception stage, because all of the background work and unconscious thought that marked the previous phase are missing.

This stage is marked by your realization that there is something to your idea. You have done enough research to discover that you could theoretically make the technology work. You've fleshed out the concept to the extent that a tentative application can be determined, although the use that you will put this technology to is still speculative at best. You're beginning to think, "I can make it work, and if it works, I might be able to use it like this."

Although the insight achieved in the "Eureka!" moment might make you think that you are finished, you have a lot of hard work remaining before the infant technology matures to the point of usefulness.

Childhood

The childhood stage marks the beginning of the true research and development effort that will lead, if successful, to the introduction of a new technology into a new or existing product. You've had your insight. Now you must verify the theory and demonstrate that you can make it work.

Much of the work that goes on during this aptly named life cycle stage appears to the outsider as playing around in the laboratory. Many researchers also refer to this work as "playing in the sand box." Your goal is to try every possible combination and permutation of the lab set up in an effort to get the experiment to work.

Finally you get the technology to work. At least you can make it work in a highly controlled laboratory environment. Once it starts to work, no matter how feebly, you'll try to force ever more performance out of the system. This happy childhood ends when someone, usually in management, finds out about your nascent technology, and you now lose control of your experiment because it's about to be released as a new product. There are still some problems that need to be ironed out, but you have proved that the concept is feasible. Not only does the technology work, but you have shown that it can work in its intended application.

Adolescence

Since you've been working so hard to get your technology to function at all, user convenience is not yet there. Still, you're finally ready to release the first version of your new technology product. This is a major happening in the life of a technology. In the Department of Defense (DoD), it marks the transition from the science and technology (S&T) laboratory to a product centered acquisition environment. In the commercial world, the transition is from the laboratory to a marketing oriented first-revenues environment.

At this stage of the technology life cycle, only nerdy early adopters will be using your product. They'll be comfortable with your technology in spite of, or maybe because of, its unfriendliness. Marketing is done mainly by word-of-mouth as one user after another discovers your product and tells friends, co-workers, and acquaintances about its cool features. More and more people begin to use the product. The increased demand forces you to improve the manufacturing process so that reliable and consistent products can be produced.

As the early adopters use the technology, they'll uncover the bugs and defects that you will have to fix before you can move on to the next phase. Toward the end of adolescence, a new class of technically unsophisticated users starts to demand new features to make your technology more user-friendly in its operation.

When the early adopters have been working with your new technology for a while, they have paved the way for a larger group of mainstream users. The first users have helped you work out the bugs, fix the defects, and promote a user friendly way to apply the technology. It is now mature enough that everyone can use it. Now it's up to marketing to turn your great technology into a great product.

Adulthood

If the marketing folks are successful, your technology will be integrated into its intended application so well that it becomes the preferred way of doing "it," whatever it is. As time passes, more and more suppliers arrive on the scene to take advantage of the booming market. Users have choices, but all of their choices are different applications of your underlying technology.

Methods of increasing market share change during this phase. At the beginning of the adult phase, the market is growing. It's possible to attract new users to a new product. This is the competitive phase, where market share, price, and brand recognition are more important than the capability of the product, since every supplier can provide roughly the same capability.

Maturity

Later, the market becomes saturated. Then competition becomes cut-throat. Everybody who needs one already has one, so the only way to increase market share is to lure customers away from the competition.

Since the market is no longer growing, suppliers try to attract customers by offering new features and upgrades. In many cases, the new features have very little to do with the application that your technology was originally intended to perform, but they do serve to differentiate one product from another. When complete market saturation has occurred, your technology moves to the next phase.

Old Age

Your technology is now old hat. Even the laggards have come to realize that this is the right way to do things. Everyone does it this way, with the possible exception of some geeky nerds who are struggling to find a new and improved method. Since your technology has achieved complete market penetration, there are no more new customers to be found. The only market is those users who need to replace worn out or broken products.

In this declining market, there is no room for new suppliers. Because even established suppliers have a hard time operating in this environment, some leave for more lucrative opportunities elsewhere. Many products start to disappear. The market becomes the preserve of a few well-established providers who cater to the demands of a steadily diminishing number of remaining customers.

Senility

Your technology has been around so long that it's starting to wear out. You've been putting patches and upgrades on it, but now you've fixed the initial technology so many times that it's hardly recognizable. As you add more fixes, the technology actually starts to decline in usefulness. It's nearing the end of its life cycle.

This is the time when some producers quit manufacturing products based on your technology, turning their attention to more profitable ventures. For a while, this may result in a profitable market for those firms that survive as they cater to the needs of a declining niche market. Individuals who stubbornly hang on to the products of this technology discover that support and spare parts become harder and harder to find.

Death

Finally, support dries up altogether, and the technology dies. This is the stage where your technology has reached obsolescence. You no longer support it. Even third party suppliers to a niche market of the few remaining customers find that it no longer pays to support your technology with materials or know-how, so they move on to other ventures or suffer bankruptcy. Your once proud technology, even with all its patches, upgrades, and fixes, can no longer keep up with newer replacement technologies. In some cases, a dead technology disappears entirely. It ends up on the scrap heap of history, possibly maintaining a tenuous existence as a museum piece or some other relic of a bye-gone era.

LET ME TELL YOU ABOUT THE WHALE

Figure 2.1 shows you what the technology life cycle looks like. In the diagram, the horizontal axis represents the passage of time going from left to right, while the vertical axis shows usefulness, or utility. The graph displays the amount of utility present at a particular point in time. As a technology matures, the chart shows how utility increases for a time, then levels off before going into a decline and eventual obsolescence. Note the slow start during the conception phase. Utility doesn't really start to take off until the technology reaches its adolescence. The rapid growth in utility flattens out during the adult, maturity, and old age stages before starting a decline to obscurity.

The Whale Chart—Did I Ever Tell You About the Whale?

Can you see why I call it, "The Whale Chart?"

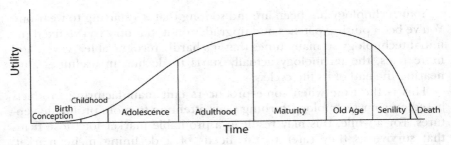

Figure 2.1.   Technology Life Cycle.

Summary

This concludes my introduction to the technology life cycle. After a discussion of some alternative life cycle descriptions, we'll tie all of the various concepts together by showing how each of them maps to the whale chart.

ALTERNATIVE LIFE CYCLES

This section illustrates that there are alternative definitions of the technology and product life cycles. I make no claim to completeness. I'm sure that you can easily find many additional life cycle definitions in the literature; however, the examples I'm providing here should convince you that there is no single approved set of life cycle definitions. All of them are roughly analogous to the biological model presented on the whale chart.

The differences are caused by emphasizing different aspects of the life cycle. If your emphasis is on product marketing, you will have a different life cycle view from that of someone who emphasizes technology development. We'll show below how focusing on different aspects of the maturity landscape lead authorities to varying, but not necessarily antagonistic, views of technology maturity.

DoD Acquisition Life Cycle

The DoD emphasizes the acquisition process in its definition of the technology life cycle. This acquisition emphasis puts a strong bias on the development end of the technology life cycle. While this acquisition emphasis appears to be reasonable, because DoD spends enormous sums of money in purchasing new weapon systems, it ignores the fact that most of a new system's life cycle costs will occur during the Operations and Support phase. This tendency is growing as weapon systems remain in use for longer and longer periods of time.

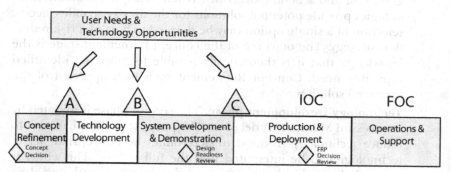

Figure 2.2.   DoD Acquisition Cycle.

The DoD life cycle displayed in Figure 2.2 above has five stages:

1.   Concept Refinement.
2.   Technology Development.
3.   System Development and Demonstration.
4.   Production and Deployment.
5.   Operations and Support.

The first four of these five stages deal with technology and product development. Only the last stage, Operations and Support, has to do with the system after it has been fielded, even though this stage represents the majority of the actual time a system must be managed. All of the other stages actually occur during the time that a new technology is undergoing research and development, then the productizing of the technology, and finally the manufacture and distribution of the technology in product or weapon system form. Let's examine each of these five stages in greater detail, using the official DoD Glossary of Defense Acquisition Acronyms and Terms as our source of information.

1. **Concept Refinement phase.** The purpose of the first phase is to refine the technology concept documented and to prepare a Technology Development Strategy. The decision to begin Concept Refinement does not constitute program initiation of a new acquisition program.

    In this phase, decision makers choose a particular technology or non-technical concept to satisfy a user's requirements. If a technology alternative is chosen, that technology often requires further maturing before it can be produced and integrated successfully into a weapon system. This could be the cause of acquisition program cost and schedule overruns. When multiple candidate technologies provide potential solutions for the user's capability needs, selection of a single option may be delayed as part of a risk reduction strategy. The outcome of the concept refinement stage is the knowledge that it is theoretically feasible to meet the identified capability need. Concept Refinement ends with approval of the preferred solution.

2. **Technology Development phase.** The second phase is initiated by a successful Milestone A decision. The purpose of this phase is it to reduce technology risk and to determine the appropriate set of technologies to be integrated into the full system. This effort is normally funded only for advanced development work and does not mean that a new acquisition program has been initiated.

    During this phase a selected technology alternative is developed and matured. Since the primary purpose of this phase is risk reduction, a final determination of which technologies to include in a weapon system acquisition program may be deferred until this phase.

3. **System Development and Demonstration phase.** Milestone B marks the formal start of a DoD acquisition program. This phase consists of two efforts, System Integration and System Demonstration. Milestone B can place the program in either one.

A program planning to proceed into System Demonstration at the conclusion of System Integration will first undergo a Design Readiness Review to confirm that the program is progressing satisfactorily during the phase.

The phase involves mature system development, integration and demonstration to support Milestone C decisions. It also supports Live Fire Test and Evaluation and Initial Operational Test and Evaluation of production representative articles.

4. **Production and Deployment phase.** The fourth phase of the life cycle consists of two efforts, Low Rate Initial Production and Full Rate Production (FRP) and Deployment. It begins after a Milestone C review. The purpose of this phase is to achieve an initial operational capability (IOC) that satisfies the mission need.

5. **Operations and Support phase.** This phase consists of Sustainment and Disposal. The phase is not initiated by a formal milestone, but instead begins with the deployment of the first system to the field. This act initiates the Sustainment effort which overlaps the Full Rate Production and Deployment effort of the previous phase. Full operational capability (FOC) is reached during the Operations and Support phase.

Did I Ever Tell You About the Whale?

If we add the DoD Acquisition Life Cycle to the whale chart shown earlier, it looks like Figure 2.3:

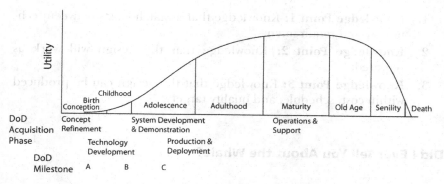

Figure 2.3. Technology Life Cycle with the DoD Acquisition Cycle.

Note that the first four phases of the acquisition life cycle are pretty far to the left. The DoD phase boundaries, marked by the decision milestones, are ideally located at the whale chart points shown, but the specific time when a milestone is reached will vary from program to program. The Government Accountability Office (GAO) reports that DoD typically moves the milestone decisions even further to the left. GAO says that initiating acquisition programs with immature technology is a major cause of cost and schedule overruns.

Concept Refinement maps to the Conception and Birth stages of the whale chart. The Technology Development phase is associated with Childhood. System Development and Demonstration usually begins in late Childhood, running over into Adolescence. Production and Deployment at Milestone C roughly maps to late Adolescence or the beginning of Adulthood. The largest part of the Whale Chart is associated with the Operations and Support phase of the DoD Acquisition Cycle.

Government Accountability Office Knowledge Points

Because such a great deal of money is spent on weapon system acquisition, the GAO also emphasizes the acquisition portion of the technology or product life cycle. In a 1998 report, GAO (then known as the General Accounting Office) identified technology maturity with three points where specific knowledge about the product (in this case, weapon system) has become known to the development team. In a burst of bureaucratic creativity, they called these points Knowledge Point 1, Knowledge Point 2, and Knowledge Point 3. The three knowledge points are defined as follows:

1. **Knowledge Point 1:** Knowledge that a match exists between technology and requirements.
2. **Knowledge Point 2:** Knowledge that the design will work as required.
3. **Knowledge Point 3:** Knowledge that the design can be produced within cost, schedule, and quality targets.

Did I Ever Tell You About the Whale?

Figure 2.4 shows us that all three knowledge points occur at the front end of the technology life cycle.

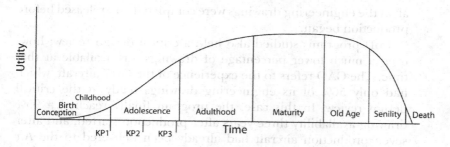

Figure 2.4. Technology Life Cycle with GAO Knowledge Points (KPs).

1. **Knowledge Point 1.** At this point, you know for sure that the capabilities of the available technology match the needs of the user. When you reach this knowledge point, the technology you plan to use is no longer a laboratory curiosity. The technology is well understood, and the manufacturing processes for producing it at reasonable cost with an acceptable level of quality exist.

    Commercial firms typically reach this knowledge point before they launch a new product development program. In this way, they isolate product development from the technology development process with all of its uncertainties.

    DoD, on the other hand, often does not reach this knowledge point until late in weapon system/product development. Often attainment of this level of knowledge does not occur until production items are available for operational testing. At that late time, if a test article fails, it could require that the weapon system program redesign portions of the system while it is already in production. Design changes during production are always far more expensive than changes carried out earlier in the development process.

2. **Knowledge Point 2.** This knowledge point is reached when approximately 90% of the engineering drawings are completed and released to manufacturing. At that time, the design is frozen, and the drawings reflect the program manager's confidence that the design will function properly as required.

    Again, commercial firms achieve Knowledge Point 2 earlier than the DoD does, in most cases. The companies included in the GAO study usually had at least 90% of the engineering drawings available for review when they performed the critical design review. This review took place about half way through product development, and was completed well before the product began production. Virtually all of the engineering drawings were completed and released before production began.

DoD programs studied also held a critical design review; however, a much lower percentage of drawings was available at that time. The GAO refers to the experience of the C-17 aircraft, which had only 56% of its engineering drawings ready at the critical design review. In this case, the program finally reached a 95% drawing availability three years after production started, and after seven production aircraft had already been delivered to the Air Force.

3. **Knowledge Point 3.** Attaining Knowledge Point 3 means that all of the key manufacturing processes are under statistical control. This is a concept borrowed from quality assurance. It means that valid data concerning the manufacturing process exists, and that the data show that the manufacturing process is capable of producing the product at an acceptable level of quality and at a reasonable cost. For some products, this implies achieving and measuring an acceptable yield. For all products, it means that the process is producing consistently high quality output at the required volume.

Commercial firms rely on statistical process control of existing manufacturing processes to achieve this knowledge point. Most commercial firms will not begin production with unproven manufacturing processes.

DoD programs often begin production before key manufacturing processes reach a state of statistical control. The GAO reports that seven years after production started, not all C-17 manufacturing processes were yet under control. There are some reasons for this. Many programs receive their money only one year at a time. Many new DoD weapon systems push the state of the art in manufacturing, but the DoD program managers don't have the money they need to develop the manufacturing process along with the new technologies. They have found that demonstrating performance is the best way to justify continuing the program, so production development takes a back seat to performance improvement.

Did I Ever Tell You About the Whale Tail?

Zoom in on the tail of the whale chart and you'll see the difference in timing of program launch and production start within commercial firms in general as contrasted with DoD programs reported by the GAO.

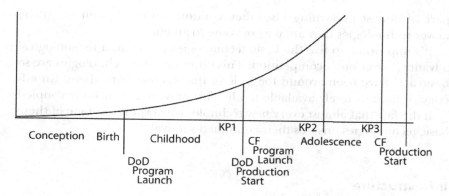

Figure 2.5. Knowledge Points (KPs) versus Program and Production Start for Commercial (CF) and DoD.

Figure 2.5 displays what the GAO reported above. DoD pushes program launch and production start so far to the left that they have not, in general, attained the knowledge needed to assure success when they make key decisions. Why does DoD rush to production before the key knowledge points have been reached? The answer is simple. It has to do with money. Acquisition programs have much more money available than S&T programs have. This forces much of the technology maturation process in DoD to take place within an acquisition program rather than within the S&T area.

EXCEPTIONS

While most technologies can be expected to follow the technology life cycle as outlined above, some fortunate technologies survive seemingly forever. These fortunate few avoid the decline, phase-out, and death portions of the Whale Chart. We'll look at two categories of technology whose members enjoy immunity from technology death.

The Basic Technologies

The first category of technologies that can live forever are the basic technologies. By basic technologies, I mean those that are included in almost all other technologies. These are such things as the simple machines like the pulley, wheel and axle, or inclined plane that have been around for as long as anyone can recall. Since these basic technologies are part of almost everything else, they continue to exist, even when the newer technologies they are part of come to an end.

It's impossible to use the basic technologies to obtain a technological advantage over one's competition. This is because the technologies are so basic and have been around for so long that no one owns them. Knowledge of them is freely available to all. This universal availability coupled with the fact that almost every new technology incorporates some of these basic technologies, insures their continued survival.

Infrastructure

While the basic technologies are pretty well defined from antiquity, new technologies can also escape the fate of obsolescence and death by becoming part of the infrastructure. When technologies achieves infrastructure status, in other words, when they become part of the normal structure of doing business, they lose their competitive advantage. It isn't that they are no longer necessary; it's just that everyone has them. The business that possesses these infrastructure technologies no longer enjoys a competitive edge; it is merely on par with everyone else. The business that does not have them, however, is at a real disadvantage.

When a technology reaches this degree of ubiquity, it has achieved the status of a commodity, like electricity or the telephone. As other technologies come and go, the infrastructure supporting all of them goes on and on. A technology that becomes an infrastructure technology may be able to beat the odds and avoid decline, phase-out, and death because it has become a necessity of business life.

I should mention that even infrastructure technologies can eventually die. It is possible for a technology to become such an integral part of the business environment that it evolves into an infrastructure commodity, yet eventually it becomes obsolete, overthrown by a new breakthrough technology.

Did I Ever Tell You About the Whale? Exceptions

This last view of the technology life cycle chart (Figure 2.6) demonstrates that basic technologies and infrastructure technologies can continue in the Maturity stage forever. I guess I shouldn't really call this chart a Whale Chart since it lost the head.

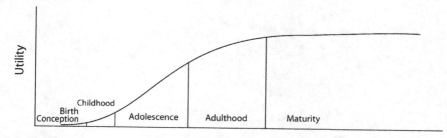

Figure 2.6. Basic and Infrastructure Technologies.

CONCLUSION

In this chapter, you learned about the technology life cycle. You saw its graphic portrayal in the Whale Chart, the central figure in much of what follows. You also found that infrastructure technologies and the basic technologies do not follow the technology life cycle.

Now that you know about the technology life cycle and its limitations, you have a good foundation for understanding technology maturity in general. The next chapter contrasts the technology life cycle with the product life cycle. There you'll see how the whale chart can be used in a whole new way to describe the life of a single product.

© 2006 M. Engling

© 2006

THE PRODUCT LIFE CYCLE

CHAPTER OVERVIEW

In this chapter, I'll again tell you about the whale, this time by describing the product life cycle in terms of the whale chart. First I'll show you the technology adoption life cycle popularized by Geoffrey Moore in his book, *Crossing the Chasm*. I'll superimpose Moore's life cycle onto the whale chart. Then I'll discuss the product life cycle. Finally, I'll show you the relationship between the technology and product life cycles by updating the whale chart.

MOORE'S TECHNOLOGY ADOPTION LIFE CYCLE

Geoffrey Moore distinguishes between successive groups of technology adopters. When plotted against time, the number of technology adopters

Did I Ever Tell You About the Whale? Or Measuring Technology Maturity, pp. 35–43
Copyright © 2008 by Information Age Publishing

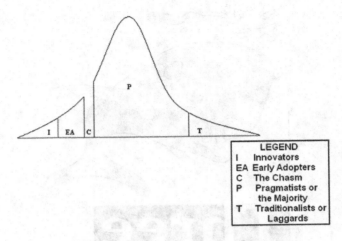

Figure 3.1.   Moore's Technology Adoption Cycle.

follows the normal bell curve. Figure 3.1 above shows Moore's technology adoption cycle.

Moore had a marketing perspective. He showed how the discontinuities at certain positions in the life cycle require your marketing strategy to change. A chasm, indicated by the letter C in the figure, separates two segments of the curve. Moore showed how your failure to account for different styles of customer behavior on either side of the chasm could result in a marketplace failure. His insights helped define market strategy appropriate to the type of technology user you were trying to attract at a given stage of the Technology Adoption Life Cycle.

- Moore divides his curve into four segments, each characterized by the type of technology adopter predominant in that segment. The four types of adopters, listed with the letter key that matches those in the diagram, are:

- I—Innovators. These are the technology enthusiasts who like technology for its own sake. They'll buy a new high-technology product because it's new, not because they need it. Innovators don't require user-friendly interfaces. In fact, they may find difficulty of use to be an attractive feature, since it separates the technically savvy users (them) from the masses.

- EA—Early adopters. These are the people who recognize that the new technology will give them some sort of competitive advantage. Unlike the innovators, early adopters use a new technology only if they can adapt it to an opportunity that is important to them. They are the visionaries who bring new methods to the marketplace. Early adopters put the new technology product to use. They're the ones who force suppliers to fix any defects and make the product easy for everyone to use.

- P—Pragmatists. Pragmatists account for almost two-thirds of the technology's adopters. They can be further divided into two subcategories, early majority and late majority. Early majority pragmatists are people who do not like to take risks, but see the advantages of using tested technologies. They are the beginning of a mass market. Late majority pragmatists dislike innovation. They believe in tradition rather than progress. They'll buy high-technology products reluctantly when they have to, but don't expect to like them.

- T—Traditionalists or laggards. Traditionalists do not like high technology products at all. They will eventually buy these products, but only when no alternative remains. By the time the laggards start buying a technology product, the innovators, and maybe some early adopters, are already enjoying the next generation product.

Although Moore calls his diagram a technology adoption life cycle, his work has a product or marketing focus. He does not start from a technology viewpoint. His adoption curve assumes that a relatively mature technology exists, a technology with at least with enough maturity to make a product that people can use.

DID I EVER TELL YOU ABOUT THE WHALE? TECHNOLOGY LIFE CYCLE AND MOORE TOGETHER

Figure 3.2 superimposes Moore's technology adoption cycle onto the technology life cycle whale chart. This chart shows that Moore's cycle begins only when a technology has reached the adolescence stage. A technology must attain some usefulness, or it can't be put into a product for people to use.

PRODUCT LIFE CYCLE

The product life cycle is concerned with the production and marketing of a product that uses existing technology in its design. The product life cycle has not been very widely used within the Department of Defense

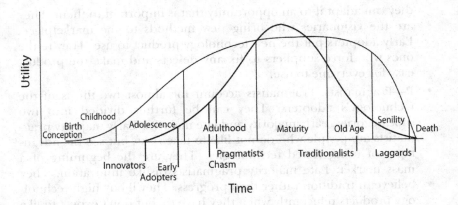

Figure 3.2.   Moore's Adoption Cycle and the Technology Life Cycle.

(DoD), probably because marketing is not much of an issue. A commercial business has a strong product or marketing focus, in most cases. This is because a commercial firm profits by selling a product. The marketing emphasis was highlighted by Peter Drucker, who wrote, "The business has two—and only these two—basic functions: marketing and innovation. Marketing and innovation produce results; all the rest are costs."

In addition to the marketing focus, the product life cycle differs from the technology life cycle in a temporal way. The product life cycle is actually embedded within the technology life cycle. This is because a technology has to exist before it can be included in a product, so the technology life cycle begins earlier than a corresponding product life cycle. A product that becomes obsolete can be replaced by a new product that still makes use of the same underlying technology. In this case, the technology life cycle continues after an individual product's life cycle has terminated.

I'm going to use a five-stage product life cycle. Each of these stages could be divided further, depending on the needs of the particular product or analysis. The five stages are:

1.   Development.
2.   Introduction.
3.   Growth.
4.   Maturity.
5.   Decline.

Development

Development here refers to product development, not technology development. The entire product development process is often called, "Productization." The name reflects the fact that this process takes an existing technology and turns it into a useful product. As such, it occurs at the tail end of the technology development cycle, but early in a technology's useful life. During the product development stage, there are no sales. This is a period of investment for possible future gain. Since resources are being consumed to create the product, but sales are non-existent, the development stage is a time of negative revenue.

Introduction

Product launch, or the introduction of a new product to the market, is an important step, because it marks the first time that the product provides revenues. Product launch is typically a huge marketing and advertising affair, accompanied by a lot of hoopla and media attention. This marketing extravaganza can be quite expensive, so the producer fervently hopes that the time spent in the introduction stage will be mercifully short. A product in this stage is not yet showing a profit, however, because first revenues are not high enough to offset the costs associated with product launch. If the product does not achieve market acceptance, it could die in this stage. For this reason, it can be risky to use brand new products as system components.

Growth

The name says it all. In this stage, growth is what's happening. As the product gains market acceptance, sales take off. The growth stage can be divided into two separate phases.

Phase I Growth

Demand will start growing at an increasing rate. In this expansion phase, it's hard not to make a profit because the market absorbs any product manufactured. Producers concentrate on improving their production capacity to meet increasing demand. This is the stage where capital investment in new production facilities or equipment can pay handsome returns, because the eager market snaps up whatever comes out of the factory.

Phase II Growth

There comes a time when the market, while still growing in absolute terms, is growing at a decreasing rate. While new customers are still out there, it's becoming increasingly difficult to find them. Toward the end of this phase, a market shakedown often occurs as less competitive companies lose market share to the leaders, and either move on to more lucrative endeavors or go out of business.

Maturity

Sales reach their peak, then level off. Prices are competitive. Customers display some brand loyalty. The product design stabilizes, and there are enough producers available so that supply is not a problem. Products in this stage present few risks to designers who use them as components in system design.

Saturation

Some would call this a separate stage, but I'm including it as part of the maturity stage. The market becomes saturated. Everyone who wants one already has one. Sales of a product are restricted to replacing items that have either worn out or otherwise have lost utility. The market becomes cut-throat as competitors try to lure customers away from each other.

Decline

When a product enters this stage, it's hard to attract new customers. Many customers are moving on to newer products, leaving a steadily diminishing customer base behind. Producers and suppliers abandon previously profitable product lines to go after new markets. Finally, a few specialty providers service the niche market that remains. As the decline progresses, it can become difficult to locate suppliers of the once plentiful product. DoD calls this the period of diminishing manufacturing sources. Manufacturing technology (MANTECH) programs attempt to alleviate problems caused when older weapon systems require spare parts that have become scarce due to obsolescence.

Phase Out

This is sometimes listed as a separate stage, not part of the decline stage. The product is now approaching the end of its useful life. The point of discontinued manufacture marks the end of the Phase Out step. This point is defined as the point of obsolescence.

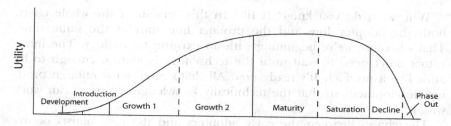

Figure 3.3. Product Life Cycle.

Did I Ever Tell You About the Whale? The Product Life Cycle

This is the product life cycle (Figure 3.3 above). Note that the form is identical to that of the whale chart used above to illustrate the technology life cycle. The difference is that here we are looking at the life cycle as it involves a single product that uses a technology which should be already available when product development first starts.

Did I Ever Tell You about the Whale? Product and Moore Together

I showed you earlier how Moore's technology adoption life cycle looks when it's placed on top of the technology life cycle whale chart. Now I want to show you how Moore's life cycle diagram fits onto the product life cycle whale chart.

I said that Moore started from a product or marketing viewpoint, not a technology perspective. If I'm right, that means the Moore diagram should fit onto the product life cycle better than it fit on the technology life cycle. Let's see how it looks (Figure 3.4).

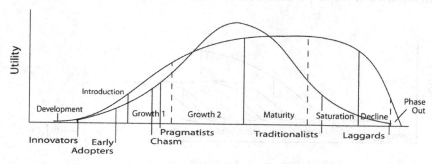

Figure 3.4. Moore's Adoption Cycle and the Product Life Cycle.

Well, what do you know! It fits! In this version of the whale chart, both the adopter line and the product line start at the same time. That's because we're beginning with an existing technology. The innovators don't need to wait until the technology is mature enough to be added to a product. It's ready now. All that's required is enough product development so that the technically knowledgeable users can work with it.

The chasm between the early adopters and the pragmatists occurs right where you'd expect it to, during growth phase 1. This is the time when all of the adventurous customers have bought the product, and you'll now have to switch your marketing appeal to target the mainstream buyers.

Did I Ever Tell You About the Whale? Product and Technology Life Cycles Together

If I put the technology life cycle together with the product life cycle, you'll see what I meant when I said that the product life cycle is embedded within the technology life cycle. The similarity in form that we noted above is seen to be a similarity in the sense of the fractals of chaos theory, with the same basic form repeated within the technology life cycle, but at a reduced scale.

Once a technology reaches late childhood or early adolescence, it will be ready to be included in a product. That's when the product life cycle first starts. As the technology matures, more and more products arise that are either based on it or incorporate it in their design. Some of these products will be replacements for earlier products that have become obsolete. Some will be upgrades and problem fixes. Others may be cosmetic "improvements" made in an attempt to prolong the life cycle of a particular product or product line. In general, the technology life cycle will outlast any single product based on the underlying technology.

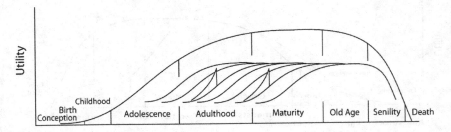

Figure 3.5.   Product and Technology Life Cycle.

In addition to the typical product life cycle curves discussed above, the diagram also includes two highly peaked life cycle curves. These represent fad products, items that quickly achieve a dominating market position only to fall even more rapidly out of fashion once the fad has run its course.

CONCLUSION

In this chapter, we took a detailed look at the product life cycle. Once more, I told you about the whale by using the Whale Chart as a picture of the product life cycle. Finally I showed how the Whale Chart can be used to demonstrate the relationship of the product life cycle to the technology life cycle.

The next chapter gets into the NASA developed Technology Readiness Levels (TRLs) that have proved to be useful for determining the current maturity of a given technology.

© 2006 M. Engling

© 2006

Four

TECHNOLOGY
READINESS LEVELS

CHAPTER OVERVIEW

One of my purposes in writing this book is to propose methods of measuring technology maturity. This chapter begins that effort. It is devoted entirely to the NASA Technology Readiness Level (TRL) scale. This is a good place to start your study of technology maturity measurement, because TRLs were one of the earliest attempts to quantify the measurement of technology maturity. After a brief introduction and historical review, you'll examine the current official TRL definitions. I'll also share my thoughts about these definitions. Then I'll show you some methods for computing TRLs. You'll finish the chapter by seeing what's good and what's not so good about using TRLs as a measure of technology maturity. Along the way, you'll discover how TRLs fit onto the whale chart.

INTRODUCING TRLs

The TRL is an attempt to measure the maturity of a technology. Technology maturity begins with science and ends with engineering. The TRL scale shows this changing nature of a maturing technology. The most recent TRL scale consists of nine different levels. The first three are generally science while the last three are almost all engineering. The middle three TRLs form a fuzzy boundary between scientific research and engineering development. As a technology moves through the nine TRL levels, it progresses from a scientific idea characterized by pure unconstrained thought (Einstein's "Gedanken experiments") to a fully developed application that has demonstrated its usefulness in a real-world operational environment.

TRL HISTORY

There seems to be some confusion about the origin of the TRL concept. Everyone agrees that the concept originated in NASA, but most articles on the subject say something like, "has been used in NASA space planning for many years." Perhaps influenced by a NASA white paper written by Mankins in 1995, the UK Ministry of Defense (MOD) dated TRLs from 1995. The Department of Defense (DoD) says, "Using TRLs to describe the maturity of technologies considered for a new system originated with NASA in the early 1980s."

I have found the NASA paper that documents the origin of the TRL scale. Writing in 1989, Stanley Sadin and his co-authors stated that the Readiness Levels concept was formulated following an analysis of historical records going back to the 1960s. Sadin's 1989 NASA paper was the first published description of Readiness Levels, although NASA had used the concept for some time before the paper was published. The early nature of this work can be deduced from the fact that the paper talks about "technology levels," "technology readiness," and "readiness levels" but does not put the terms together into "technology readiness levels." The abbreviation "TRL" is never used in the Sadin paper. There are no references to TRLs before his paper.

Sadin referred to seven levels of technology readiness, defined in Table 4.1.

The Sadin group showed how the seven readiness levels incorporate the more traditional breakdown of research and development into four loosely defined categories:

Table 4.1   Readiness Levels

Level	Description
Level 1	Basic Principles Observed and Reported
Level 2	Potential Application Validated
Level 3	Proof of Concept Demonstrated Analytically and/or Experimentally
Level 4	Component and/or Breadboard Laboratory Validated
Level 5	Component and/or Breadboard Validated in Simulated or Real-Space Environment
Level 6	System Adequacy Validated in Simulated Environment
Level 7	System Adequacy Validated in Space

- Basic research
- Feasibility
- Development
- Demonstration

Anticipating the conclusions of a 1999 Government Accountability Office (GAO) study, Sadin stated "When historical records are analyzed, it can be demonstrated that the difference between success and failure is traceable to the adequacy, or depth, of the (Advanced Research and Technology) program in pursuing technology readiness." The paper goes on to say that use of the seven levels can help eliminate ambiguous understandings and misconceptions of the planned depth of technology programs between researchers, managers, and technology users.

In their 1999 report, the GAO, like Sadin before them, concluded that "The experiences of DoD and commercial technology development cases GAO reviewed indicate that demonstrating a high level of maturity before new technologies are incorporated into product development programs puts those programs in a better position to succeed." GAO went on to encourage the use of "a disciplined and knowledge-based approach of assessing technology maturity, such as TRLs, DoD-wide."

The GAO gave the use of TRLs in DoD a tremendous boost when they recommended their use in weapon system acquisition. Although a few DoD organizations, such as the Air Force Research Laboratory (AFRL), had been using the TRL scale in a few programs, it was only after the GAO report came out that DoD began a real drive to use TRLs routinely. Technology Readiness Assessments are now required at both Milestones B and C. The DoD Acquisition Guidebook recommends the use of TRLs when managers perform a Technology Readiness Assessment.

The use of TRLs has also gone international. The UK MOD recommended their use in a TRL Guide published in 2002. The Canadian armed forces have also used TRLs to measure technology maturity.

TRL DEFINITIONS

The TRL definitions currently accepted by DoD follow a standard format. Each TRL has a number, a brief definition, and a description. The thermometer chart (Figure 4.1) shows the number and the definition for the TRLs. For TRL 1, the TRL definition is, "Basic Principles Observed and Reported." The TRL description is the subject of the next section.

The thermometer diagram (Figure 4.1) gives the current NASA definitions. The thermometer chart also displays an overlapping grouping of TRLs to show how, as the TRL number increases, the technology matures from basic research through technology development and demonstration to system operation. For the first seven TRLs, the grouping closely follows Sadin's (Table 4.1) four-category grouping (basic research, feasibility, development, and demonstration), although the category labels used on the chart have changed slightly.

If you look closely at the thermometer chart (Figure 4.1), you'll notice that TRLs 1 through 7 are almost identical with Sadin's 7 readiness levels

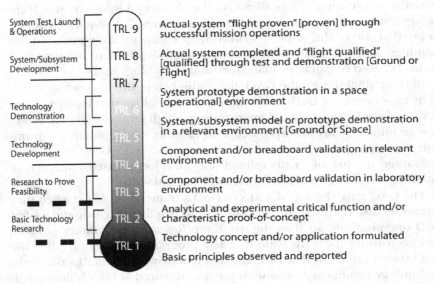

Figure 4.1.   Technology Readiness Levels [When there is a difference between the NASA and DoD definitions, the DoD verbiage is given in square brackets.]

(Table 4.1). TRLs 8 and 9 extend the readiness level concept through flight qualification to actual mission operations.

TRL DEFINITIONS, DESCRIPTIONS, SUPPORTING INFORMATION, AND COMMENTS

The TRL descriptions amplify the rather sparse definitions given above to provide more detail. The descriptions try to explain what each definition means. I suppose you could think of a description as a definition of a definition. Take a look at the following example:

TRL 2, Technology Concept and/or Application Formulated ——— Description
Invention begins. Once basic principles are observed, practical applications can be invented. Applications are speculative and there may be no proof or detailed analysis to support the assumptions. Examples are limited to analytic studies.

Publications or other references that outline the application being ——— Supporting Information
considered and that provide analysis to support the concept.

You can see that the description gives you more information than you found in the definition. However, the description doesn't seem to be complete. You might wonder what the words "Applications are speculative" mean. You could also try to figure out why the description says, "There may be no proof or detailed analysis," and follows this statement with, "Examples are limited to analytic studies."

To help you understand the requirements for achieving each TRL, the Department of Defense *Technology Readiness Assessment (TRA) Deskbook* gives you a list of supporting information for each level. The supporting information provides the evidence that a specific technology has reached the level claimed.

The official TRL definition and description system ends with these four elements: TRL number, definition, description, and supporting information. Obviously, these elements don't tell the entire story. More information could make the TRLs more readable, maybe more understandable, and perhaps even more useful. The information in the "**Comment**" section at the end of each TRL entry gives you my perspective and interpretation of what the TRL means.

TRL 1, Basic Principles Observed and Reported

Lowest level of technology readiness. Scientific research begins to be translated into applied research and development. Examples might include paper studies of a technology's basic properties.

Supporting Information: Published research that identifies the principles that underlie this technology. References to who, where, when.

Comment: At this level of understanding, you are first beginning to recognize the scientific principles involved. Your experimental work in the laboratory consists mostly of observations of the physical world. Experiments are designed to obtain new knowledge through these observations by applying the scientific method. You're not worried about whether or not the science can be applied to a real world problem. When you have achieved TRL 1, you will have limited evidence that your observations are repeatable and dependable. You'll use these observations to derive the underlying scientific principles that could form the basis of a new or improved technology.

TRL 2, Technology Concept and/or Application Formulated

Invention begins. Once basic principles are observed, practical applications can be invented. Applications are speculative and there may be no proof or detailed analysis to support the assumptions. Examples are limited to analytic studies.

Supporting Information: Publications or other references that outline the application being considered and that provide analysis to support the concept.

Comment: When you go from TRL 1 to TRL 2, you're moving from pure research to applied research. The emphasis here is still on understanding the science, but you're beginning to think about possible applications of the scientific principles. You are doing mostly analytical or paper studies and view graph engineering. You are still doing some experimental laboratory work, but the purpose of this work is to corroborate the basic scientific observations made during TRL 1. You are not trying to get the technology to work. In fact, at this level, you don't yet have a new technology. You're still trying to understand the science better. The step between TRL 1 and 2 consists of knowing the physics well enough to conclude that there may be some actual uses for the scientific principles, although such technological uses are not yet well defined. You're finally able to say, "If it works, we might be able to use it like this."

The scientific output from both TRL 1 and 2 is mostly scientific papers and journal articles. You don't yet have any hardware to show off your technology. The newborn technology is no more than "vaporware" at these early maturity levels.

TRL 3, Analytical and Experimental Critical Functions and/or Characteristic Proof-of-Concept

Active research and development is initiated. This includes analytical studies and laboratory studies to validate physically analytical predictions of separate elements of the technology. Examples include components that are not yet integrated or representative.

Supporting Information: Results of laboratory tests performed to measure parameters of interest and comparison to analytical predictions for critical subsystems. References to who, where, and when these tests and comparisons were performed.

Comment: At TRL 3, you're moving beyond the paper phase. You perform laboratory experiments, not to discover new scientific principles, but to see whether you can verify that the concept does what you expect it to do in a predictable, repeatable manner. If the concept proves to be feasible, you can start working on individual elements or components that could eventually form a functioning system. You are performing laboratory work to validate pieces of the technology without trying to integrate components into a complete system. Modeling and simulation may be used to complement physical experiments. Until now, your emphasis has been on validating the predictions made during earlier analytical studies so that you're certain that the technology concept has a firm scientific underpinning. Now you're ready to move on to developing a specific application of the science.

TRL 4, Component and/or Breadboard Validation in Laboratory Environment

Basic technological components are integrated to establish that they will work together. This is relatively "low fidelity" compared to the eventual system. Examples include integration of "ad hoc" hardware in the laboratory.

Supporting Information: System concepts that have been considered and results from testing laboratory-scale breadboard(s). References to who did this work and when. Provide an estimate of how breadboard hardware and test results differ from the expected system goals.

Comment: Look at TRLs 4 through 6 as the bridge from scientific research to engineering. At TRL 4, you'll start to put individual components together to see whether you can get them to work as a system. At the previous level, you proved that you understood the science well enough to get the individual technology components to work. Now you're trying to see whether you can get them to work together in a laboratory breadboard system that might resemble the final system in function only. You'll put the components into this early laboratory system to see whether the concept is feasible. The components used in the design are not the components that you would use for an operational system. The laboratory mock-up you put together is a kludge of on-hand lab assets and some special purpose devices that may require special handling, calibration, or alignment to get them to function. Your purpose is to increase your understanding of the technology, demonstrating that it has the potential to be useful in an application. You're not ready to test it in a real world operating environment yet. You're still struggling to get it to work reliably in the laboratory.

TRL 5, Component and/or Breadboard Validation in Relevant Environment

Fidelity of breadboard technology increases significantly. The basic technological components are integrated with reasonably realistic supporting elements so they can be tested in a simulated environment. Examples include "high fidelity" laboratory integration of components.

Supporting Information: Results from testing a laboratory breadboard system are integrated with other supporting elements in a simulated operational environment. How does the "relevant environment" differ from the expected operational environment? How do the test results compare with expectations? What problems, if any, were encountered? Was the breadboard system refined to more nearly match the expected system goals?

Comment: This TRL is distinguished by increases in the accuracy of the controlled environment in which the technology is tested. The big difference between TRL 4 and TRL 5 is the realism of the working environment. This is where the nascent technology is first exposed to an operating environment that approximates what it will encounter outside the laboratory. The controlled environment of the laboratory is now modified to be more like the expected operational environment. As the TRL definition says, the technology is being validated in a relevant environment that simulates key aspects of the operational environment. This is not necessarily a highly stressing environment. After all, you're still in the laboratory.

At TRL 5, the breadboard of TRL 4 becomes a brass board by improving the fidelity of the individual components and interfaces. While your experimental set-up does not yet reach the quality of a prototype, it's a lot better then the laboratory kludge of TRL 4. This TRL marks a substantial increase in the fidelity of both the system, including its components and interfaces, and the operating environment it is exposed to.

TRL 6, System/Subsystem Model or Prototype Demonstration in a Relevant Environment

Representative model or prototype system, which is well beyond that of TRL 5, is tested in a relevant environment. Represents a major step up in a technology's demonstrated readiness. Examples include testing a prototype in a high fidelity laboratory environment or in simulated operational environment.

Supporting Information: Results from laboratory testing of a prototype system that is near the desired configuration in terms of performance, weight, and volume. How did the test environment differ from the operational environment? Who performed the tests? How did the test compare with expectations? What problems, if any, were encountered? What are/were the plans, options, or actions to resolve problems before moving to the next level?

Comment: TRL 6 begins true engineering development of the technology as an operational system as opposed to development of an abstract technology with possible application to real world problems. The brass board mock-up of the previous TRL is improved so that it is now a functional prototype. It represents the full system in function but not necessarily in form. Your prototype is capable of performing all of the functions that will be required of the operational system. It is not yet packaged so that it will physically fit into the intended platform. You have yet to demonstrate its performance across the entire operational environment, although you are moving out of controlled laboratory conditions.

TRL 7, System Prototype Demonstration in an Operational Environment

Prototype near or at planned operational system. Represents a major step up from TRL 6 by requiring demonstration of an actual system prototype in an operational environment (e.g., in aircraft, in a vehicle, or in space). Examples include testing the prototype in a test bed aircraft.

Supporting Information: Results from testing a prototype system in an operational environment. Who performed the tests? How did the test compare with expectations? What problems, if any, were encountered?

What are/were the plans, options, or actions to resolve problems before moving to the next level?

Comment: The system prototype improves to the point where you can use it as a pre-production prototype. Almost all of the system elements and components are available. You know how the components will fit together. All of the physical and functional interfaces are clearly defined. By the end of TRL 7, the engineering design should be essentially frozen. From now on, you should expect only minor design changes. Although the system may not yet be installed on the intended weapon system or platform, the prototype will be exposed to the actual operational environment on a surrogate platform or test bed.

Technologies used in space systems may not always follow this step-by-step progression from one TRL to the next. In some cases, it will not be possible to design a "relevant environment" in an earthbound laboratory. In these cases, a technology may have to move directly from TRL 6 to TRL 9 without going through the intervening steps. Such technologies will almost always exhibit a high probability of failure when first deployed.

TRL 8, Actual System Completed and Qualified Through Test and Demonstration.

Technology has been proven to work in its final form and under expected conditions. In almost all cases, this TRL represents the end of true system development. Examples include developmental test and evaluation of the system in its intended weapon system to determine if it meets design specifications.

Supporting Information: Results of testing the system in its final configuration under the expected range of environmental conditions in which it will be expected to operate. Assessment of whether it will meet its operational requirements. What problems, if any, were encountered? What are/were the plans, options, or actions to resolve problems before finalizing the design?

Comment: You are done with system development. You have an actual production quality system. If necessary, developmental test and evaluation (DT&E) has been accomplished to demonstrate that the production system meets all design requirements. Any design changes you make at this level will be the minimum required to bring the system up to specification. You will not be changing any system performance parameters, interfaces, or configuration requirements. You're ready to make a production decision.

TRL 9, Actual System Proven Through Successful Mission Operations

Actual application of the technology in its final form and under mission conditions, such as those encountered in operational test and evaluation (OT&E). Examples include using the system under operational mission conditions.

Supporting Information: OT&E reports.

Comment: TRL 9 means that you have an operational system. Any required OT&E has been accomplished. The production system is either ready for deployment or has already been deployed to operational units. If OT&E is not a program requirement, a technology can achieve TRL 9 by performing a live operational mission.

DID I EVER TELL YOU ABOUT THE WHALE? TRLs

When TRLs are added to the Whale Chart (Figure 4.2), you'll notice that all 9 TRLs occur early in the technology life cycle. This is because the TRL focuses on technology development. Once a technology has been successfully incorporated in an actual product and used on an operational mission (DoD and NASA) or for its intended purpose (commercial market), it has reached TRL 9. This is the end of the TRL scale. This development point normally occurs during the life cycle phase called "Adolescence," although in some cases it could happen earlier. One thing is certain. By the time a new technology enters the adult phase, it has gone beyond the realm measured by the TRL.

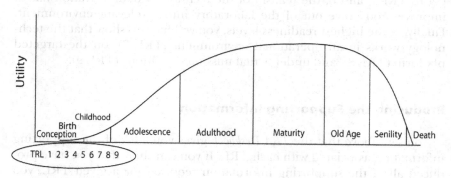

Figure 4.2.   Technology Life Cycle with TRLs.

CALCULATING THE TECHNOLOGY READINESS LEVEL

You now know the nine TRLs, but I haven't shown you how to figure out what TRL your technology has achieved. There are several ways to approach this task. I'll show you four of them: applying the definitions, looking at the technology demonstration environment, producing the supporting information, and using the AFRL TRL Calculator.

Applying the Definitions

You could determine your technology's maturity level by looking at the TRL definitions and deciding which one best matches the current state of your technology's development. This method is simple and straightforward; however, it is not very objective. TRL definitions are open to different interpretations. As you learned when I discussed the TRL definitions, the basic definitions are augmented by descriptions and supporting information. These additions were made because the TRL definitions alone don't give you enough information to come up with a meaningful technology readiness assessment.

Technology Demonstration Environment

At a high level, you can think of the TRL as a description of the realism of the environment in which the technology has been demonstrated. At the lowest readiness levels, TRLs 1 through 3, you have formulated the technology concept in an academic environment, but you haven't actually demonstrated the technology and its application to a real world problem. At TRL 4, you will show that you can get it to work in your laboratory. To get to TRL 5 and 6, the realism of the technology demonstration has to improve. You move out of the laboratory into a relevant environment. Finally, at the highest readiness levels, you will have to show that the technology works in an operational environment (TRL 7), on the targeted platform (TRL 8), and under actual mission conditions (TRL 9).

Producing the Supporting Information

You saw above that the *TRA Deskbook* gives you a list of the supporting information associated with each TRL. If you can show that you have produced all of the supporting information required for a given TRL, you will have made a strong case that you have achieved that level. This is the mandated method for supporting a technology readiness assessment at the DoD level. However, even if you have all of the required supporting information available, it is still possible for you to overlook some essential steps in developing and maturing your technology.

TRL Calculator

The AFRL TRL Calculator automates the process of computing the technology's maturity by applying an engineering "bottom-up" approach. The calculator is a Microsoft Excel® spreadsheet that asks you a series of questions, shoves bamboo splinters under your fingernails until you provide acceptable responses,[1] and computes the TRL achieved based on your answers. The latest version of the TRL Calculator includes questions that cover the supporting information required by the *TRA Deskbook*. Because you have to answer each individual question at every readiness level, the TRL Calculator keeps you from overlooking necessary steps in maturing your technology. I included a copy of the AFRL TRL Calculator with the companion software that comes with this book. The software also contains program documentation for the calculator.

WHAT ARE TRLs GOOD FOR?

TRLs give you a snapshot of a technology's maturity at a given instant. They provide you with a quick description of the current state of a technology. Because of this, TRLs can be used to benchmark technology maturity. You can use the current TRL as an indication of how much development work you have already done.

TRLs can also serve as a communication device. In those cases where a technology developer must hand off a technology to another organization or corporate entity for product or system development, TRLs can help both sides understand exactly what each is required to do. By providing a common reference point for the technology developer and user, TRLs can help to eliminate misunderstandings and ambiguities in the technology transition process. TRLs can form a part of a technology's exit criteria. The exit criteria tell the technology developer that the development program has met its goals.

In their 1999 report, the GAO showed that TRLs can be used as an indicator of a program's risk. They argued, like Sadin had argued earlier, that when a technology has sufficiently matured in the laboratory as a science and technology (S&T) program before it is incorporated into an acquisition program, risk is decreased. Figure 4.3, adapted from the GAO report, shows how the amount of information that is unknown about a technology decreases as TRLs increase. According to the GAO, once a technology reaches a maturity of approximately TRL 7, its risk for transition goes from high to low. This means that a more mature technology is less likely to encounter unanticipated problems in cost, schedule, or performance when added to a weapon system than a less mature technology.

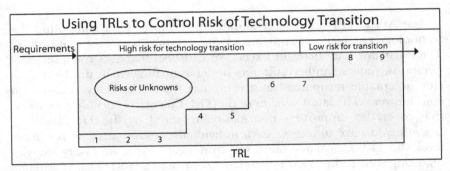

Figure 4.3.   TRLs as a Measure of Program Risk.

TRLs tell us what has been done in a technology development program up to a certain point in time. As such, they can provide an indication of the relative degree of risk assumed by incorporating the technology into a product or system development program at that time. TRLs can also be used to measure progress towards a defined technology maturity goal.

WHAT ARE TRLs NOT GOOD FOR?

Since TRLs give a historic picture of the current state of a technology's maturity, their predictive value is low. Knowing the current TRL of a technology doesn't tell anything about how difficult it will be to achieve the next TRL. You know what's been done up until now. You might have a good idea of what you still have left to do. What's missing is knowledge of how hard the next step or steps will be.

We said above that TRLs provide an indicator of the remaining degree of risk in a program. This is true in a relative sense only. TRLs will not help you identify specific risks as part of a risk management program. All you learn from the TRL is the technology maturity. You know that a more mature technology carries less development risk than a less mature technology does. You just don't know what the risk might be, either for the entire program risk or the risk involved in reaching the next TRL.

You saw above in the discussion of TRLs and the whale chart that all nine TRLs occur early in the technology life cycle. TRLs do not provide any information about a technology's maturity for most of its life. In a touch of irony, TRLs become useless once the technology they're measuring has matured to the point of usefulness. You cannot use TRLs to determine the coming obsolescence of a particular technology because the TRL scale terminates when a technology matures. For the same reason, TRLs won't help you identify problems of reduced manufacturing availability toward the end of the technology life cycle.

TRLs describe increasing levels of technological maturity as a technology progresses from an initial idea to a fully capable product. In assessing the maturity of a technology, it is important to identify any technology advance needed by the new concept or application. If there is no technological advancement required, the underlying technology is mature and the concept has achieved a high TRL rating.

TRLs do not give you a full picture of a technology's maturity. You saw in Chapter 1 that technology maturity is a multi-dimensional concept. Since the TRL measures only one aspect of the technical dimension, it cannot give complete information. The Software Engineering Institute (SEI) reports that TRLs give the program managers of advanced technology development efforts information on only approximately 30% of the factors they need to follow.

The SEI team also found that TRLs were used mainly to improve the timing of inserting a technology into a product development program. The TRL metric appears to have its greatest utility as part of the exit criteria for a S&T program. Other potential uses in program management and risk management have so far not been fully realized.

CONCLUSION

This chapter introduced the TRL as a measure of technology maturity. I talked about the history of TRLs and discussed the nine levels in some detail. I told you about the whale, putting TRLs onto the technology life cycle whale chart. I also talked about some uses and limitations of this metric. Finally, I introduced you to the TRL Calculator, a tool you can use to compute the TRL for your technology.

The next chapter extends the TRL concept into the world of software development. There you will find a set of coherent TRLs for software intensive technologies.

NOTE

1. The bit about the bamboo splinters is an outright lie.

© 2006 M. Engling

© 2006

SOFTWARE TECHNOLOGY READINESS LEVELS

CHAPTER OVERVIEW

You will now see how to extend the Technology Readiness Level (TRL) concept to software. I'll first discuss the idea of software maturity. Then I'll introduce you to the history of software TRLs. I'll also give you my opinion on which of the available software TRL sets is the best. Finally, I'll show you several proposed versions of software TRL descriptions including my comments on them.

BUT SOFTWARE IS DIFFERENT...

Software and hardware are different. This "startling" conclusion leads you to wonder whether a concept such as technology maturity can be applied

Did I Ever Tell You About the Whale? Or Measuring Technology Maturity, pp. 61–76
Copyright © 2008 by Information Age Publishing
All rights of reproduction in any form reserved.

to software. After all, software doesn't wear out. So what could you possibly mean if you were to say that software ages? How can software go through the stages of the technology life cycle discussed earlier? It would appear that software either exists, or it does not exist. If it exists, it won't wear out, because it has no moving parts, so how could it age?

One way you can see aging, or growth, in software technology is in its development. A new software capability does not spring into existence fully formed. In its development, software goes through the same basic steps we talked about earlier. First, someone gets an idea for a new capability that could be realized in software. This new capability could be one that has always been done by physical hardware components, but now someone thinks that he or she can get it done by properly programming the capability into a software package.

While the idea remains in the, "Wouldn't it be nice if we could get software to do this?" stage, you're in the conception phase. As soon as you have the "Ah, ha!" experience that says, "This is the way I might be able to do it," you've gone on to the birth phase. When you play around with the idea, trying to prove that the concept might really work, your software technology is in its childhood. It moves into adolescence when it has acquired enough usefulness to actually do something. It's probably not very user friendly yet, but it can serve a purpose for those adventurous enough to work through its limitations.

When your software technology becomes an accepted way of doing something, it's moved into adulthood. If it passes on to the state where it is THE way of doing something, your software has reached maturity. Up until now, your software has been steadily increasing in utility as it matures. This increase in utility comes from all the bug fixes, patches, and upgrades that you have been adding.

At some point in its life cycle, your software actually begins to decline in usefulness and quality. Eventually, it will reach the point where all of the additional code you added to fix bugs or to increase usefulness causes software problems, such as hang-ups or slow-downs. This is a sign of old age. When the problems become severe, you may see signs of senility in your once robust software product. Senility occurs because the changes you have made overwhelm the software architecture. The period of senility is characterized by an actual decline in product utility. Finally, your now obsolete software product fades from use and dies.

Did I Ever Tell You About the Whale?

Software technology fits neatly onto the whale chart and into the biological technology maturity model. That's hardly surprising, because the whale chart was originally produced by a team[1] working on software maturity. While the group did not use the biological technology life cycle model, they did recognize that software can eventually decline in utility as it becomes senile. The team also noted that TRLs occur at the front end of the software technology life cycle. This fact makes the TRL ineffective as a measure of maturity once a product enters service.

The whale chart developed by the software maturity working group looks just like the one we've been using, but with different divisions and nomenclature. Unlike the technology life cycle whale chart, this model stretches out the TRLs further along the horizontal axis. When looking at software TRLs, the team took a product life cycle view rather than a technology life cycle view. The product development section of this chart goes all the way into the adult phase of the biological model, while we cut off the technology development section during the adolescent phase in the earlier chapter.

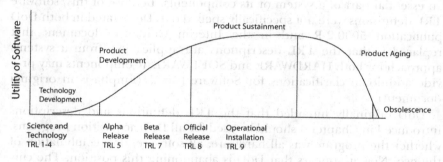

Figure 5.1. Notional Software Product Maturity Life Cycle.

HISTORICAL BACKGROUND

NASA

Besides having pride of place in the invention and first use of TRLs, NASA also pioneered the use of TRLs for software. The earliest set of software TRLs that I found was the set reported in a NASA presentation in 1999.[2] This set of software TRLs follows the nine level format of the earlier hardware TRLs. NASA created the precedent of maintaining the original nine TRL definitions while altering the descriptions to fit the software development environment.

The NASA descriptions split the nine TRLs into three groups of three. The first three TRLs demonstrate the scientific feasibility of the software technology. TRLs 4 through 6 are concerned with showing that the technology can be implemented using known software engineering principles and processes. The final three levels deal with system implementation and integration, culminating in an operational software system that performs in the operational environment.

NASA has recently developed a new set of software TRL descriptions. Their new set of software TRLs has suggested exit criteria listed for each TRL. The exit criteria give you the tasks that must be completed before you can say that you have achieved each level. I have included the new NASA information on Table 5.1, TRL Table for Software.

Department of Defense

When the Department of Defense (DoD) first directed the use of TRLs, the same TRL definitions were applied to both hardware and software. The TRLs and descriptions of the previous chapter included software as an essential part of a system or its components. Because of this, software TRL definitions were not specifically spelled out. DoD stated in both DoD publication 5000.2-R and in the Interim Guidance document that replaced it that the TRL descriptions are applicable, "from a systems approach for both HARDWARE and SOFTWARE. (Components may provide additional clarifications for Software [sic])." (Emphasis in original document).

DoD originally intended that the TRL definitions and descriptions introduced in Chapter 4 should be used for all DoD acquisition programs, whether the program was all hardware, all software, or a combination of the two. Now it appears that DoD is abandoning this position. The current on-line version of DoD's Technology Readiness Assessment (TRA) Deskbook provides a new set of definitions and descriptions for software TRLs.

Army

A U.S. Army team under Dr. John Niemela of the Communications-Electronics Command's (CECOM) Research, Development and Engineering Center (CERDEC)[3] was the first DoD component to publish TRL "clarifications" for software as permitted by the DoD guidance quoted above. Dr. Larry Stotts, the U.S. Army's Director of Technology, asked Dr. Niemela, "Are TRLs the right measurement for software?" The team soon realized that the accepted TRL definitions were difficult to apply to software intensive systems. They created a set of TRLs for software compatible with the existing hardware TRLs and delivered the results to Dr. Stotts by October 2000.

The genius of the Army team showed itself in the way they adapted the accepted TRL descriptions to software. The team used the terminology of the DoD descriptions, but reworked them to show a software perspective. You can find these Army descriptions in Table 5.1.

Missile Defense Agency

The Missile Defense Agency (MDA) took a different approach to software maturity measurement. Instead of basing their measure on the TRL, they used a variation of the five-step Engineering and Manufacturing Readiness Level (EMRL) scale. The five Software Readiness Levels (SWRLs) are made up of a combination of attributes from four different software aspects. The four aspect categories are Algorithms and Functions, Architecture and Environments, Software Technical Maturity, and Software Process Progress.

Adding up the contributions from each aspect will give an overall rating of SWRL 1 through SWRL 5. The higher the number, the greater the maturity of the software technology. Greater maturity implies increased readiness to be adopted into a software or hardware and software system. From lowest to highest, the five SWRLs were defined by Rob Gold of the MDA as follows:

SWRL 1—Concept. Alternatives, including possible software reuse are explored. It may be possible to meet the system's requirements by creating or modifying a software solution using available software development processes.

SWRL 2—Prototype. Models and analysis have been performed to develop high level architecture, interface descriptions, and algorithms and functions necessary to create the software solution.

SWRL 3—Development. Software requirements are finalized, functional architecture created, interface requirements flow down through a work breakdown structure. Key algorithms have been selected and documented, and coding begins with formal reviews.

SWRL 4—Functional. Coded functions run on the target hardware in a simulated operational environment. Testing may include hardware in the loop or computer in the loop simulations.

SWRL 5—Deployable. Software system has fully documented capability to operate in the fielded system as demonstrated by deployment testing or actual mission operations.

While this five step SWRL system served its intended purpose for the MDA when it first was published, it is now being modified to conform more closely with the TRL nine-level standard. You should note that the SWRL measures the maturity of a software product, not a software technology. Even though SWRLs as now proposed are moving toward a nine-level format, their product orientation means that the two measures are not interchangeable, although they are related.

AFRL

I started working on TRLs for software in late 2002 when I found that the TRL descriptions for software intensive systems left something to be desired. Looking over the proposed Army clarifications, I thought that I could see some discrepancies between them and the basic NASA/DoD TRL descriptions. Then I discovered the NASA Software TRLs. I thought that they, too, had some shortcomings. In my opinion, these two software TRL versions were not internally consistent. They didn't always seem to follow the maturity progression defined by the hardware descriptions. Also, neither of them enjoyed universal acceptance. For these reasons, I proposed my own set of draft software TRL descriptions. I combined all four sources (DoD, Army, NASA, and my own ideas) into one coherent set of Software TRL descriptions. I used them in some of my earlier work, notably the 2002 release of the TRL Calculator.

DoD Information Technology Working Group

In November 2004, an Information Technology Working Group (ITWG) tackled the software TRL issue. The Department of Defense TRA Deskbook provides the set of Software TRL definitions and descriptions this working group developed along with a list of supporting information that should be used at each TRL to show achievement of that level.

When I reviewed the work of the ITWG, I realized that they had duplicated the work I had attempted with my draft set of software TRL descriptions. While there are differences between the two sets of descriptions, these differences are relatively minor. Since the DoD TRA Deskbook includes the ITWG work, I believe that set ought to be adopted as the standard set of software TRL descriptions within DoD and possibly across the wider federal government and commercial community interested in software technology maturity measurement.

INTRODUCTION TO TRL TABLE FOR SOFTWARE

In Table 5.1, I'll start each readiness level with the DoD TRL definition. From TRL 4 on, I'll also give you the ITWG definitions for software. (DoD and ITWG definitions are identical for TRL 1 through TRL 3). Next, I'll provide the software TRL descriptions from the DoD, NASA, and Army sources mentioned above. I'll put all three descriptions together side-by-side so you can see how they fit, and also where they don't fit together very well.

Beneath these three descriptions I'll show the ITWG description and supplemental information requirement taken from the DoD TRA Handbook. This set of software TRL descriptions is the set that I want everyone to adopt.

I'll follow the ITWG data with the latest NASA software TRL descriptions and exit criteria. By putting the NASA data here, I will maintain the chronological order of the information on the table.

Finally, I'll present my comments about each readiness level and tell you what I think each one really means. Where necessary, I'll also comment on differences between and among the several software TRL descriptions.

When you see all of the candidate software TRLs together, you should be able to see why I favor the widespread adoption of the ITWG set. The ITWG software TRLs capture the similarities among the predecessor software descriptions. In fact, even the later NASA work seems to be included in the ITWG work at almost every TRL. I find that this is true not only in the NASA and ITWG TRL descriptions, but also in the NASA exit criteria and ITWG supporting information requirements.

CONCLUSION

This concludes my discussion of applying TRLs to software. As you have seen, it can be done. Bear in mind, however, that the same limitations on the use of TRLs apply to software as to hardware. You learned in Chapter 4 that TRLs do not give you a full picture of a technology's maturity. The predictive power of software TRLs is also poor. While the TRL can tell you the software technology's current maturity, it will not warn you that your product is in danger of becoming obsolete. Like the hardware TRL, the TRL for software is useless at the end of the technology life cycle.

In the next chapter I'll discuss some other methods of measuring technology maturity. Most of them are based on the TRL concept.

Table 5.1. Software TRLs

TRL 1: Basic Principles Observed and Reported

DoD—Lowest level of technology readiness. Scientific research begins to be translated into applied research and development. Examples might include paper studies of a technology's basic properties.	**NASA 1999**—Basic properties of algorithms, representations, and concepts. Mathematical formulations. Mix of basic and applied research.	**Army**—Lowest level of software readiness. Basic research begins to be translated into applied research and development. Examples might include a concept that can be implemented in software or analytic studies of an algorithm's basic properties.
ITWG—Lowest level of software technology readiness. A new software domain is being investigated by the basic research community. This level extends to the development of basic use, basic properties of software architecture, mathematical formulations, and general algorithms.		**Supplemental Information**—Basic research activities, research articles, peer-reviewed white papers, point papers, early lab model of basic concept may be useful for substantiating the TRL level.
NASA 2006—Scientific knowledge generated underpinning basic properties of software architecture and mathematical formulation.		**Exit Criteria**—Peer reviewed publication of research underlying the proposed concept/application.
My Comments—While we may have some idea of what we're trying to do, we really haven't thought it through yet. The possibilities are still speculative, although it looks like there may be something there. We're answering the question, "Wouldn't it be nice if we could get software to do this?"		

TRL 2: Technology Concept and/or Application Formulated

DoD—Invention begins. Once basic principles are observed, practical applications can be invented. Applications are speculative and there may be no proof or detailed analysis to support the assumptions. Examples are limited to analytic studies.	**NASA 1999**—Basic principles coded. Experiments with synthetic data. Mostly applied research.	**Army**—Invention begins. Once basic principles are observed, practical applications can be invented. Applications are speculative and there is no proof or detailed analysis to support the assumptions. Examples are limited to analytic studies.
ITWG—Once basic principles are observed, practical applications can be invented. Applications are speculative, and there may be no proof or detailed analysis to support the assumptions. Examples are limited to analytic studies using synthetic data.		**Supplemental Information**—Applied research activities, analytic studies, small code units, and papers comparing competing technologies.

NASA 2006—Invention begins. Practical application exists but is speculative. No experimental proof or detailed analysis is available to support the conjecture. Underlying algorithms are created and documented.

My Comments—If it works, we can apply it like this. We're starting to think of potential applications for the new software technology. We may have a specific research and development (R&D) program in mind. We have done some preliminary analysis, but we still haven't proved that the concept will work. We do know what major tasks and functions are required, but not yet at the module or function point level of detail.

Exit Criteria—Documented description of the application / concept that addresses feasibility and benefit.

TRL 3: Analytical and Experimental Critical Function and/or Characteristic Proof-of-Concept

DoD—Active research and development is initiated. This includes analytical studies and laboratory studies to physically validate analytical predictions of separate elements of the technology. Examples include components that are not yet integrated or representative.

NASA 1999—Limited functionality implementations. Experiments with small representative data sets. Scientific feasibility fully demonstrated.

Army—Active research and development is initiated. This includes analytical studies to produce code that validates analytical predictions of separate software elements. Examples include software components that are not yet integrated or representative but satisfy an operational need. Algorithms run on a surrogate processor in a laboratory environment.

ITWG—Active R&D is initiated. The level at which scientific feasibility is demonstrated through analytical and laboratory studies. This level extends to the development of limited functionality environments to validate critical properties and analytical predictions using nonintegrated software components and partially representative data.

NASA 2006—Development of limited functionality to validate critical properties and predictions using non-integrated software components

Supplemental Information—Algorithms run on a surrogate processor in a laboratory environment, instrumented components operating in laboratory environment, laboratory results showing validation of critical properties.

Exit Criteria—Documented analytical/experimental results validating predictions of key parameters

My Comments—We've done enough analysis to show that the software can be made to work. We have worked out the algorithms for the major functions identified at TRL2, maybe doing some exploratory coding to get some of the more speculative ideas to run on a laboratory computer. There is no attempt to integrate functions and/or databases. Input and output (I/O) is done manually, or in a "brute force" manner. We're looking around to see if we can reuse or adapt existing software.

Table continues on next page.

Table 5.1. Continued

TRL 4: DoD—Component and/or Breadboard Validation in Laboratory Environment
ITWG—Module and/or Subsystem Validation in a Laboratory Environment (i.e., Software Prototype Development Environment)

DoD—Basic technological components are integrated to establish that they will work together. This is relatively "low fidelity" compared to the eventual system. Examples include integration of "ad hoc" hardware in the laboratory.

NASA 1999—Stand alone prototype implementations. Experiments with full-scale problems or data sets.

Army—Basic software components are integrated to establish that they will work together. They are relatively primitive with regard to efficiency and reliability compared to the eventual system. System software architecture development initiated to include interoperability, reliability, maintainability, extensibility, scalability and security issues. Software integrated with simulated current /legacy elements as appropriate.

ITWG—Basic software components are integrated to establish that they will work together. They are relatively primitive with regard to efficiency and robustness compared with the eventual system. Architecture development initiated to include interoperability, reliability, maintainability, extensibility, scalability, and security issues. Emulation with current/legacy elements as appropriate. Prototypes developed to demonstrate different aspects of eventual system.

Supplemental Information—Advanced technology development, stand-alone prototype solving a synthetic full-scale problem, or standalone prototype processing fully representative data sets.

NASA 2006—Key, functionally critical, software components are integrated, and functionally validated, to establish interoperability and begin architecture development. Relevant environments defined and performance in this environment predicted.

Exit Criteria—Documented test performance demonstrating agreement with analytical predictions. Documented definition of relevant environment.

My Comments—This level of technology development is primarily concerned with the coding of individual modules and functions. We do some ad hoc integration, but we're still force-feeding the output of one module into another with little regard for the final interface characteristics. We know enough about the software project to do initial estimates of software size for program risk management and cost estimation.

TRL 5: DoD—Component and/or Breadboard Validation in Relevant Environment
ITWG—Module and/or Subsystem Validation in a Relevant Environment

DoD—Fidelity of breadboard technology increases significantly. The basic technological components are integrated with reasonably realistic supporting elements so it can be tested in a simulated environment. Examples include "high fidelity" laboratory integration of components.

ITWG—Level at which software technology is ready to start integration with existing systems. The prototype implementations conform to target environment/interfaces. Experiments with realistic problems. Simulated interfaces to existing systems. System software architecture established. Algorithms run on a processor(s) with characteristics expected in the operational environment.

NASA 2006—End to end software elements implemented and interfaced with existing systems conforming to target environment, including the target software environment. End to end software system tested in relevant environment meets predicted performance. Operational environment performance predicted.

NASA 1999—Prototype implementations conform to target environment/interfaces. Experiments with realistic problems. Simulated interfaces to existing systems.

Army—Reliability of software ensemble increases significantly. The basic software components are integrated with reasonably realistic supporting elements so that it can be tested in a simulated environment. Examples include "high fidelity" laboratory integration of software components. System software architecture established. Algorithms run on a processor(s) with characteristics expected in the operational environment. Software releases are "Alpha" versions and configuration control initiated. Verification, Validation and Accreditation (VV&A) initiated.

Supplemental Information—System architecture diagram around technology element with critical performance requirements defined. Processor selection analysis, Simulation / Stimulation Laboratory buildup plan. Software placed under configuration management. COTS/GOTS in the system software architecture are identified.

Exit Criteria—Documented test performance demonstrating agreement with analytical predictions. Documented definition of scaling requirements.

My Comments—We're not ready to do any releases yet, not even at the Alpha level. We're still in a laboratory environment, although the processor may be representative of the target application. Architecture is well defined with laboratory level integration of software modules and functions. Configuration management scheme and test protocol are documented.

Table continues on next page.

Table 5.1. Continued

TRL 6: DoD—System/Subsystem Model or Prototype Demonstration in a Relevant Environment
ITWG—Module and/or Subsystem Validation in a Relevant end-to-end Environment

DoD—Representative model or prototype system, which is well beyond that of TRL5, is tested in a relevant environment. Represents a major step up in a technology's demonstrated readiness. Examples include testing a prototype in a high fidelity laboratory environment or in simulated operational environment.

NASA 1999—Prototype implementations on full-scale realistic problems. Partially integrated with existing hardware / software systems. Limited documentation available. Engineering feasibility fully demonstrated.

Army—Representative model or prototype system, which is well beyond that of TRL5, is tested in a relevant environment. Represents a major step up in software demonstrated readiness. Examples include testing a prototype in a live/virtual experiment or in simulated operational environment. Algorithms run on processor or operational environment integrated with actual external entities. Software releases are "Beta" versions and configuration controlled. Software support structure in development. VV&A in process.

ITWG—Level at which the engineering feasibility of a software technology is demonstrated. This level extends to laboratory prototype implementations on full-scale realistic problems in which the software technology is partially integrated with existing hardware/software systems.

Supplemental Information—Results from laboratory testing of a prototype package that is near the desired configuration in terms of performance, including physical, logical, data, and security interfaces. Comparisons between tested environment and operational environment analytically understood. Analysis and test measurements quantifying contribution to system-wide requirements such as throughput, scalability, and reliability. Analysis of human-computer (user environment) begun.

Exit Criteria—Documented test performance demonstrating agreement with analytical predictions.

NASA 2006—Prototype software partially integrated with existing hardware/ software systems and demonstrated on full-scale realistic problems.

My Comments—Software is at the prototype level implying an Alpha release with the start of rigorous VV&A and configuration management. This is the first attempt to subject the system to a realistic, albeit simulated, operational environment. We can expect numerous bug fixes and upgrades as deficiencies are discovered.

TRL 7: DoD—System Prototype Demonstration in an Operational Environment
ITWG—System Prototype Demonstration in an Operational High-Fidelity Environment

DoD—Prototype near or at planned operational system. Represents a major step up from TRL 6, requiring demonstration of an actual system prototype in an operational environment, such as in aircraft, vehicle, or space. Examples include testing the prototype in a test bed aircraft.

NASA 1999—Most functionality available for demonstration and test. Well integrated with operational hardware/software systems. Most software bugs removed. Limited documentation available.

Army—Represents a major step up from TRL 6, requiring the demonstration of an actual system prototype in an operational environment, such as in a command post or air/ground vehicle. Algorithms run on processor of the operational environment integrated with actual external entities. Software support structure in place. Software releases are in distinct versions. Frequency and severity of software deficiency reports do not significantly degrade functionality or performance. VV&A completed.

ITWG—Level at which the program feasibility of a software technology is demonstrated. This level extends to operational environment prototype implementations where critical technical risk functionality is available for demonstration and a test in which the software technology is well integrated with operational hardware/software systems.

Supplemental Information—Critical technological properties are measured against requirements in a simulated operational environment.

NASA 2006—Prototype software is fully integrated with operational hardware/software systems demonstrating operational feasibility.

Exit Criteria—Documented test performance demonstrating agreement with analytical predictions.

My Comments—I think that at TRL 7 we're still dealing with a prototype system, not an operational one. Since we're still working with a software system prototype, any software releases will be configuration controlled Beta versions subject to operational (field) conditions. The verification step of VV&A is completed, demonstrating that the software prototype meets the established requirements as documented in the software system specifications.

Table continued on next page.

Table 5.1. Continued

TRL 8: DoD—Actual System "flight qualified" Through Test and Evaluation

ITWG—Actual System Completed and Mission Qualified Through Test and Demonstration in an Operational Environment

DoD—Technology has been proven to work in its final form and under expected conditions. In almost all cases, this TRL represents the end of true system development. Examples include developmental test and evaluation of the system in its intended weapon system to determine if it meets design specifications.

NASA 1999—Thoroughly debugged software. Fully integrated with operational hardware and software systems. Most user documentation, training documentation, and maintenance documentation completed. All functionality tested in simulated operational scenarios. V & V completed.

Army—Software has been demonstrated to work in its final form and under expected conditions. In most cases, this TRL represents the end of system development. Examples include test and evaluation of the software in its intended system to determine if it meets design specifications. Software releases are production versions and configuration controlled, in a secure environment. Software deficiencies are rapidly resolved through support structure.

ITWG—Level at which a software technology is fully integrated with operational hardware and software systems. Software development documentation is complete. All functionality tested in simulated and operational scenarios.

NASA 2006—The final product in its final configuration is successfully [demonstrated] through test and analysis for its intended operational environment and platform (ground, airborne or space).

Supplemental Information—Published documentation and product technology refresh build schedule. Software resource reserve measured and tracked.

Exit Criteria—Documented test performance verifying analytical predictions.

My Comments—Attainment of TRL 8 means that the software is ready to be installed in an operational system. VV&A is completed with the software accredited for use on the target weapon system. Any required Developmental Test & Evaluation has also been accomplished. The software has not yet demonstrated its functionality under actual operational mission conditions.

TRL 9: DoD—Actual System "flight proven" Through Successful Mission Operations
ITWG—Actual System Proven Through Successful Mission-Proven Operational Capabilities

DoD—Actual application of the technology in its final form and under mission conditions, such as those encountered in operational test and evaluation. Examples include using the system under operational mission conditions.

NASA 1999—Thoroughly debugged software readily repeatable. Fully integrated with operational hardware and software systems. All documentation completed. Successful operational experience. Sustaining software engineering support in place. Actual system fully demonstrated.

Army—Actual application of the software in its final form and under mission conditions, such as those encountered in operational test and evaluation. In almost all cases, this is the end of the last "bug fixing" aspects of system development. Examples include using the system under operational mission conditions. Software releases are production versions and configuration controlled. Frequency and severity of software deficiencies are at a minimum.

ITWG—Level at which a software technology is readily repeatable and reusable. The software based on the technology is fully integrated with operational hardware/software systems. All software documentation verified. Successful operational experience. Sustaining software engineering support in place. Actual system.

Supplemental Information—Production configuration management reports. Technology integrated into a reuse "wizard"; out-year funding established for support activity.

NASA 2006—The final product is successfully operated in an actual mission.

Exit Criteria—Documented mission operational results.

My Comments—A TRL of 9 means that the software system is operational. We have an operational software system that is ready for deployment or has already been deployed to operational units. Operational suitability has been demonstrated through Operational Test & Evaluation or by performing actual operational missions. TRL 9 means the software has worked at least once in an operational mission environment.

NOTES

1. Dr. Richard Turner, Department of Defense Software Intensive Systems office, chaired the team of Dr. Barry Boehm, University of Southern California; Paul Casely, UK Defence Science and Technology Laboratory; Suzanne Garcia, Software Engineering Institute; Delmar Gessner, Air Force AFOTEC; Gary Hafen, Lockheed Martin; Frank Herman, Fraunhofer Center for Experimental Software; John McGarry, Army Picatinny Arsenal, Practical Systems and Software Measurement; William Nolte, Air Force AFRL; Patricia Oberndorf, Software Engineering Institute; and Donald Reifer, University of Southern California.
2. The NASA Software definitions are included in Table 5.1.
3. Team members included Dr. Matthew Fisher, Senior Member of Technical Staff at the Software Engineering Institute; Dr. Som Karamchetty, Army Research Laboratory; Dr. Michael Macedonia, Army Simulation and Training Command; Mr. Cenap Dada of the Communications-Electronics Command's Software Engineering Center.

© 2006 M. Engling

© 2006

READINESS LEVEL PROLIFERATION

CHAPTER OVERVIEW

In this chapter, I'll introduce you to readiness level proliferation, the tendency of managers to create variants of the classic technology readiness level (TRL) to help measure and manage specific programmatic areas of concern. You'll discover proposed applications of the readiness level concept that are different from the TRLs you encountered in Chapters 4 and 5. You will find that these applications fall into two categories, "other readiness levels" and "readiness level corollaries." After I explain the differences between these two categories, I'll show you some examples of each.

Did I Ever Tell You About the Whale? Or Measuring Technology Maturity, pp. 77–91
Copyright © 2008 by Information Age Publishing
All rights of reproduction in any form reserved.

WHAT IS READINESS LEVEL PROLIFERATION?

Readiness level proliferation is my terminology for the tendency of those responsible for research and development (R&D) program oversight to add new requirements for reporting undefined readiness levels as measures of program progress. As people in upper management become aware of problems in their acquisition programs, they naturally want to prevent a recurrence of these problems. In an attempt to measure program progress in areas where they have experienced past failures, some managers impose new measurements, loosely based on the readiness level concept. They hope that, by tracking these newly prescribed readiness levels, they will be able to identify potential problems in time to take preventive management action.

WHY DOES READINESS LEVEL PROLIFERATION HAPPEN?

I believe that the perceived success of the classic TRL is responsible for the proliferation phenomenon. Managers want a quick method of measuring a program's progress. The measure is even better if it can be reported as a number. TRLs gave managers a quick, numerical measure of a technology's achieved maturity. Progress between program reviews was easily demonstrated when the TRL number increased from one review to the next. When the number didn't change, or even worse, got smaller, management attention was required. Since TRLs appeared to work well as a snapshot of technology maturity, managers expanded the readiness level concept to other portions of their programs requiring attention.

IS READINESS LEVEL PROLIFERATION A PROBLEM?

The number of proposed and in-use readiness levels seems to be growing without limit. Does this mean that readiness level proliferation is really a problem? I guess that the short answer to this question is, "It depends." When the need for a new readiness measure is real, and when there is a well-defined process for obtaining the measure, proliferation is not a problem. Properly done, a new readiness level that fills a valid need can be an effective solution to a management problem. It's only when the new readiness level is poorly thought out, or when there is no defined process for accurately measuring the level, that serious problems in readiness level proliferation occur.

The tendency to create new readiness levels probably has noble roots; however, implementation of a new measure often causes problems at the working level. Often the new readiness levels are poorly defined. Sometimes the use of a new measure is mandated before methods for computing it are developed. The reporting of these readiness levels then becomes a subjective exercise in giving the boss the number he or she is looking for. The use of a new readiness level can provide the appearance of meticulous attention to detail while underlying program management problems remain undiscovered.

My main objection to readiness level proliferation is that most of the new readiness levels should be captured in the requirements definition phase of R&D program initiation. If the requirements for such things as logistics, integration, human/system interface, etc., were properly specified up front, there would be no need to invent "readiness levels" to track them. The existing TRL methodology would capture all of these requirements since TRL is, or ought to be, based on demonstrating that a technology has met operational requirements and design specifications in a defined environment.

OTHER READINESS LEVELS VERSUS READINESS LEVEL COROLLARIES

The "Other" Category

When a readiness level is used to measure a technology's maturity along one of the multiple dimensions introduced in Chapter 1, I'm calling it one of the "other readiness levels." These readiness levels are maturity measures of the same hardware and software technologies you met in Chapter 4 and Chapter 5. They simply measure an aspect of maturity different from the pure technology focus of the TRL. Analogous readiness levels have been defined in the following areas, although only one of these has so far received wide-spread acceptance:

- Other Readiness Levels
  - Programmatic Readiness Levels.
  - Manufacturing Readiness Levels/Engineering and Manufacturing Readiness Levels.
  - Sustainment/Supportability Readiness Levels/Logistics Readiness Levels.
  - Integration Readiness Levels.

The "Corollaries" Category

Following the lead of researchers at the Carnegie-Mellon Software Engineering Institute (SEI), I'll call those applications that apply the TRL concept to different kinds of technologies "readiness level corollaries." The readiness level definitions proposed here are the same as the TRL definitions we discussed earlier. The normal TRL descriptions, however, have been altered to meet the idiosyncrasies of the specific technology.

In Chapter 5, you learned that the TRLs originally developed for hardware technologies could also be applied to software technologies. Here you'll see that the TRL concept can also be used to measure the maturity of four other types of technology in the purely technical dimension.

- Readiness Level Corollaries
  o Bio-Medical Technologies Readiness Levels.
  o Learning Systems Readiness Levels.
  o Practice-Based Technologies Readiness Levels.
  o Modeling and Simulation Readiness Levels.

Now I'll tell you about these two readiness level categories in greater detail. I'll begin with the four other readiness levels.

OTHER READINESS LEVELS

Programmatic Readiness Levels

When I first began working with TRLs in 2002, I created a spreadsheet application called the TRL Calculator. The spreadsheet included a series of questions about a technology development program. The answers were used to compute the current TRL for that program.

Users of an early version of the TRL Calculator pointed out that some of the questions I had included really had nothing to do with technology readiness. Those questions dealt with programmatic concerns, not purely technical issues. My first response was to suggest omitting those questions from the calculator. I was told that the questions were important to a technology program manager. I then invented a new category of readiness to address these programmatic concerns. At first I called the category "Program Readiness for Transition (PRT)." In later versions of the TRL Calculator, I changed the term to "Programmatic Readiness Levels (PRLs)" to align better with the TRLs.

The PRLs measure a technology's maturity in the programmatic dimension. This dimension consists of three components:

- Documentation.
- Customer focus.
- Budget.

Documentation

When I talk about documentation, I'm referring to the old adage, "The job isn't finished until the paperwork is done." The kinds of program documentation you need will vary as a program progresses through development. When your technology is in early development, say TRL 1– 3, your required documentation may only consist of journal articles or conference papers. If you're performing the work under a contract, periodic progress reports could be specified as contract deliverables. You may also be required to submit a final technical report.

As your knowledge about the technology and its potential applications grows, your paperwork needs also grow. By the time your technology has reached TRL 4 or 5, you'll need quite a bit of program documentation. Your files will bulge with things like formal requirements documents, work breakdown structures, systems engineering plans, and configuration management plans. Depending on the type of program, you could have preliminary design drawings or test and evaluation plans. If your program includes software, you will probably have done some software sizing in terms of estimates of source lines of code or number of function points. A formal list of science and technology program exit criteria, the all important factors that tell you when you're finished, will round out your program file. You may combine several of these documentation requirements into a single technology transition plan. This plan tells how and when your technology will move out of the laboratory. In the Department of Defense (DoD), transition usually means that program responsibility for further development (productization) passes from the science and technology laboratory to the acquisition community.

Once your technology reaches the maturity of TRL 6 or 7, you'll have to add a lot more documentation to your program file. You will probably have a Test and Evaluation Master Plan (TEMP), and possibly a Systems Engineering Plan (SEP) as well. You will have completed most of your hardware design drawings, at least in draft form. At this level of maturity, you'll be documenting how well your technology can meet the customer's logistics concerns as well as the technical performance requirements. If your program includes software, you will begin the formal release process. At TRL 6, your first official release will be a carefully restricted "Alpha" test release. By TRL 7, you will release a "Beta" test version to typical users. For both hardware and software systems, you will need to have a formal configuration control process in place and documented at this maturity level.

The documentation for TRL 8 and 9 centers around testing. At TRL 8, your technology will have completed its first big test, developmental test and evaluation (DT&E). This testing verifies that the system as you built it meets all of the design specifications. DT&E answers the question, "Did I build the thing right?" The DT&E test report documents completion of this testing, and is proof that your technology has achieved TRL 8. For software systems, or systems that include large amounts of software, this level of maturity also requires the completion of any necessary verification, validation, and accreditation.

TRL 9 testing is the operational test and evaluation (OT&E). This testing validates that the technology can perform its intended function under expected operational conditions. It answers the question, "Did I build the right thing?" Sometimes your technology will bypass this level of testing, and demonstrate achievement of TRL 9 by performing an actual operational mission. This is especially true for space systems, because testing in space is indistinguishable from operating in space. Because testing is, or can be, very expensive, there is a recent tendency to combine DT&E with OT&E.

Customer Focus

A good program manager always keeps his or her customer in mind. You can't keep your customer in mind if you don't know who your customer is. This is not always easy in a technology development program because, as your knowledge grows, your customer will probably change. To determine who your customer is at any time during your research and development effort, ask yourself, "If I'm successful, who will care?" The answer defines your customer or customers of the moment as your technology matures.

Keeping the customer in mind means not only knowing who your customer is but also knowing what your customer expects of you. Every time your customer changes, you'll need to revisit your customer's expectations. Even when you keep the same customer, you'll probably find that your customer's expectations and requirements change during the time it takes you to mature the technology.

When you first start to research a new technology, you won't know how it can be applied to real world problems. Your customer will not be the final user of the technology, but will be whoever is paying your salary. This customer expects you to perform scientific investigations to discover what is possible and what is not possible when working with this technology. The only output this customer wants is journal articles and a final report that document the scientific or technical knowledge you learned while performing the contracted level of work.

As your knowledge improves, you'll start to think of ways to apply the technology. The speculative applications you come up with will change the answer to the "Who will care?" question. You will now have a customer in mind who may actually be interested in putting your technology to use. You should try to get this potential customer involved as the user member of your development team. As a minimum, the user ought to work with you to create a list of operational requirements the technology must meet to fill a need. You also need to start getting involved with the acquisition office that will eventually be responsible for integrating your technology onto a new or existing product, system, or platform.

One of the last steps in obtaining customer buy-in is getting the customer to commit funds to your development effort. Once you reach this level of customer involvement, your hardest work in this area is done. When customers have put money into your project, you have guaranteed their future interest. Now "all" you have to do is deliver the technology as promised.

Budget

Compared to other programmatic dimensions, the budget has one big advantage. There is a built-in measuring stick: money. You can compare how much you have spent to your spending plan. If you have spent more than planned, you're in trouble. Running out of money means your project is cancelled.

There is a lot of material on budget management in the business literature, so I won't cover this subject in much detail. Current trends include such methods as earned value management. No matter what management technique you choose to use, you must pay attention to your budget or your project is doomed.

Manufacturing Readiness Levels

If you can't *make* it, you can't *use* it. Recognizing this fact, DoD has been working on ways to determine and report whether a technology can be manufactured using today's industrial processes and capabilities. Manufacturing or producibility maturity, as we saw in Chapter 1, is one of the multiple technology maturity dimensions. The current measure of maturity in this dimension is the Manufacturing Readiness Level (MRL). The MRL for measuring manufacturing and producibility is the one "other readiness level" that has become accepted throughout DoD. You will often find the MRL reported along with the TRL during readiness assessments and reviews.

Engineering and Manufacturing Readiness Levels

Developed by the Missile Defense Agency, Engineering and Manufacturing Readiness Levels (EMRLs) measure your capability to manufacture a technology. EMRLs capture the amount of design and manufacturing knowledge available for product development, demonstration, production, and deployment. The five EMRL levels correspond to the six TRLs running from TRL 4 through TRL 9.

Table 6.1. Engineering and Manufacturing Readiness Levels

EMRL	Definition
1	System, component or item validation in initial relevant engineering application/bread board, brass board development. Product Design/Definition phase. Technologies must have matured to at least TRL 4 or 5.
2	System, component or item in prototype demonstration beyond bread board, brass board development. Product Development Phase. Technologies must have matured to at least TRL 6 or 7.
3	System, component or item in advanced development. Product Demonstration Phase. Technologies must have matured to at least TRL 8.
4	Similar system, component or item previously produced or in production. Or, the system, component or item is in low rate initial production. Technologies must have matured to at least TRL 9.
5	System, component or item is in full rate production with continuous product/process improvement.

Source: Dr. Thomas D. Fiorino, *Engineering Manufacturing Readiness Levels (EMRLs): Fundamentals, Criteria and Metrics*, presentation to OSD/ATL Program Manager's Workshop, 3 June 2003

Manufacturing Readiness Levels

Eventually, EMRLs evolved into DoD specific MRLs which are analogous to the TRLs discussed in Chapter 4. MRLs measure both producibility of the product or system and the maturity of the manufacturing process. As presently proposed, MRLs group levels 1 through 3. This is because technologies that have not reached TRL 3 are not well enough defined to determine what manufacturing processes will be used to produce them. The MRL tracks risk elements across all the different acquisition phases, starting in R&D and culminating with an added MRL 10 to capture continuous manufacturing process improvement during technology production.

Table 6.2. DoD Manufacturing Readiness Levels (MRLs)

MRL	Definition	Description
1–3	Manufacturing concepts Identified.	Identification of current manufacturing concepts or producibility needs based on laboratory studies.
4	System, component, or item validation in a laboratory environment.	This is the lowest level of production readiness. Technologies must have matured to at least TRL 4. At this point, few requirements have been validated, and there are large numbers of engineering/design changes. Component physical and functional interfaces have not been defined. Materials, machines, and tooling have been demonstrated in a laboratory environment. Inspection and test equipment have been demonstrated in a laboratory environment. Manufacturing cost drivers are identified. Producibility assessments have been initiated.
5	System, component, or item validation in initial relevant environment. Engineering application/ breadboard, brassboard development.	Technologies must have matured to at least TRL 5. At this point, all requirements have not been validated, and there are significant engineering / design changes. Component physical and functional interfaces have not been defined. Materials, machines, and tooling have been demonstrated in a relevant manufacturing environment, but most manufacturing processes and procedures are in development (or Manufacturing Technology [MANTECH] initiatives are ongoing). Inspection and test equipment have been demonstrated in a laboratory environment. Production cost drivers/goals are analyzed. System-level design to cost (DTC) goals are set. Producibility assessments ongoing.
6	System, component or item in prototype demonstration beyond breadboard, brassboard development.	During the prototype demonstration phase, requirements are validated and defined. However, there are still many engineering / design changes, and physical and functional interfaces are not yet fully defined. Technologies must have matured to at least TRL 6. Raw materials are initially demonstrated in relevant manufacturing environment. Similar processes and procedures have been demonstrated in relevant manufacturing environment. At this point, there are likely major investments required for machines and tooling. Inspection and test equipment should be under development. Producibility risk assessments ongoing and trade studies conducted. A production Cost Reduction Plan is developed. Production goals are met.

Table continues on next page.

Table 6.2.   DoD Manufacturing Readiness Levels (MRLs)

MRL	Definition	Description
7	System, component or item in advanced development.	Technologies must have matured to at least TRL 7. At this point, engineering/design changes should decrease. Physical and functional interfaces should be clearly defined. All raw materials are in production and available to meet planned Low Rate Initial Production (LRIP) schedule. Pilot line manufacturing processes and procedures set up and under test. Processes and procedures not yet proven or under control. During this phase, initial producibility improvements should be underway. DTC estimates are less than 125 percent of goals. Detailed production estimates are established.
8	System, component or item in advanced development. Ready for LRIP.	Technologies must have matured to at least TRL 8. At this point, engineering/design changes should decrease significantly. Physical and functional interfaces should be clearly defined. All raw materials are in production and available to meet planned LRIP schedule. Manufacturing processes and procedures have been proven on the pilot line and are under control and ready for LRIP. During this phase, initial producibility risk assessments should be completed. Production cost estimates meet DTC goals.
9	System, component, or item previously produced or in production, or the system, component, or item is in LRIP. Ready for Full Rate Production (FRP).	During LRIP, all systems engineering/design requirements should be met, and there should only be minimal system engineering/design changes. Technologies must have matured to at least TRL 9. Materials are in production and available to meet planned production schedules. Manufacturing processes and procedures are established and controlled in production to three-sigma or some other appropriate quality level. Machines, tooling, and inspection and test equipment deliver three-sigma or some other appropriate quality level in production. Production risk monitoring is ongoing. LRIP actual costs meet estimates.
10	System, component, or item previously produced or in production, or the system, component, or item is in FRP.	The highest level of production readiness. Minimal engineering / design changes. System, component, or item is in production or has been produced and meets all engineering, performance, quality, and reliability requirements. All materials, manufacturing processes and procedures, and inspection and test equipment are controlled in production to six sigma or some other appropriate quality level in production. A proven, affordable product is able to meet the required schedule. Production actual costs meet estimates.

Sustainment/Supportability/Logistics Readiness Levels

Recently people in DoD have realized that emphasizing supportability during technology development is critical. You need to consider supportability early in the technology life cycle. Sustainment is an important factor in maturing a technology. In my opinion, the best way to include supportability in your R&D program is to make sure that you include supportability requirements as a subset of the program's performance requirements.

You could also measure and report the technology's sustainment or supportability readiness. You can do this through readiness level proliferation by assigning supportability readiness levels (SRLs) that are analogous to the familiar TRLs. There is, at this writing, no such measure. While the concept of the SRL exists, no one has yet created or defined such an unnecessary measuring tool.

The Navy's Aging Aircraft Integrated Product Team (IPT), however, is attempting to define logistics readiness levels (LRLs). The LRL is a new concept intended to assist the program manager in considering sustainment issues for projects that insert new technology into existing systems. In defining the LRL, the IPT wanted to provide a method and appropriate benchmarks for assessing the level of logistics support a technology has achieved at each phase of the acquisition life cycle. They also wished to provide a management tool to forecast logistics workload, manpower requirements, and to identify gaps in logistics support. Evaluation criteria are established for each project phase and used to determine LRL's for six distinct life cycle phases:

- Lab Test or R&D.
- Project Definition.
- Project Development and Implementation.
- Engineering Validation.
- Fleet Verification.
- Fleet Use.

At each of these phases, there is a benchmark set of appropriate logistics tasks. The benchmark answers the question, "What tasks must be completed at each project phase for each logistics element?" Different logistics elements will not require the same level of effort in every phase. The LRL for each phase is computed by calculating the percent of the total logistics benchmark tasks that has been completed. Each phase LRL falls into one of five levels, numbered from 0 through 4. As the percentage of completed tasks increases, the LRL also increases. The five levels are:

- LRL = 0 Unsupported (0% of required tasks completed).
- LRL = 1 Poorly Supported (1-50% of required tasks completed).
- LRL = 2 Moderately Supported (51-70% of required tasks completed).
- LRL = 3 Nearly Supported (71-99% of required tasks completed).
- LRL = 4 Fully Supported (100% of required tasks completed).

Optimally, all applicable phases would have a grade of LRL 4 if all logistics elements were addressed, resolved, or supported. The strength of this system is that the meaning of the LRL, percent of tasks complete, stays constant across all phases. The tasks included in the benchmarks change with each phase.

Integration Readiness Levels

At a recent conference, the keynote speaker stated that DoD would soon start to require program managers to report on integration readiness levels (IRLs) during acquisition milestone reviews. When I tried to find out more about this readiness level, I soon discovered that the IRL is still only a concept, and there is no real definition for this proposed measurement. Almost everyone would agree that, especially for complex systems and systems composed of other systems, integration is a serious problem. The integration problem can also be significant when you are trying to add a new technology to an existing vehicle or platform. Although the technology integration problem is present in almost every R&D program, trying to solve the integration problem by requiring the reporting of an acquisition program's achieved IRL is, to me, a quintessential example of readiness level proliferation. In fact, it was this specific case that caused me to recognize that proliferating readiness levels can be a problem.

READINESS LEVEL COROLLARIES

Bio-Medical Technology Readiness Levels

The U.S. Army Medical Research and Materiel Command drafted a set of TRL descriptions for five distinct categories of bio-medical technologies in 2001. The five categories are:

- Pharmaceutical (i.e., drugs).
- Pharmaceutical (i.e., biologics/vaccines).

- Medical Devices.
- Medical Knowledgeware.
- Medical Information Management/Information Technology & Medical Informatics.

In each of these categories, the Army used the standard TRL definitions. They not only modified the TRL descriptions, but also provided TRL Decision Criteria to help decide when you have achieved a particular level. Whenever possible, the decision criteria were based on external events, such as Food and Drug Administration approvals.

Learning Systems Technology Readiness Levels

In September 2000, an advisory panel sponsored by NASA recommended applying adaptations of TRLs to next generation learning systems. The panel's recommended learning system TRLs are shown along with the standard definitions in Table 6.3, next page. Note that the learning system TRLs stop at TRL 8. When you put the two sets of definitions side-by-side, it looks like it's actually TRL 1 that's missing in the learning system TRL list. I think that there is a good correspondence between standard TRLs 2 through 9 and learning system TRLs 1 through 8. What do you think?

Practice-Based Technologies

Several researchers from the Carnegie-Mellon University SEI have adapted the TRL concept to fit what they call "Practice-Based Technologies (PBTs)." They defined a PBT as a technology incorporating practices, processes, methods, approaches, or frameworks rather than the usual hardware and software technologies. Examples of PBTs are such things as acquisition practices, technology transition processes, and the SEI framework: Capability Maturity Model for Integration (CMMI).

A PBT does not deal with a hardware or software product, as we usually understand the term. The operating environment for PBTs is people in organizations or communities, not hardware arranged in systems. With an emphasis on organizational as opposed to technical problems, a PBT blurs the distinction between knowledge and technology. The PBT environment, with its human focus, is far more changeable, malleable, and flexible than the operating environment of purely technical or physical systems.

In adapting the TRL scale to PBTs, SEI first noted that as the hardware or system TRL progressed from 1 to 9, there was a corresponding

Table 6.3. Learning System Technology Readiness Levels

TRL	Standard Definition	Learning System
1	Basic principles observed and reported	Learning System Concept and Critical Functions Postulated and Acknowledged
2	Technology concept and/or application formulated	Critical Functions Validated (Proof-Of-Concept)
3	Analytical and experimental critical function and/or characteristic proof-of-concept	Technology Tools/Methods Validated in a Controlled Testbed Environment
4	Component and/or breadboard validation in laboratory environment	Technology Tools/Methods Validated In an Authentic Learning Environment
5	Component and/or breadboard validation in relevant environment	System Prototype Demonstrated in a Controlled Learning Environment
6	System/subsystem model or prototype demonstration in a relevant environment	System Prototype Validated in an Authentic Learning Environment
7	System prototype demonstration in a space environment	Operational System "Curriculum Qualified" Through Consecutive Test and Feedback
8	Actual system completed and "flight qualified" through test and demonstration	Operational System "Infusion Ready" Based On Authentic, Successful Curriculum Demonstrations
9	Actual system "flight proven" through successful mission operations	

Source: *Infusing Next Generation Learning Systems & Technologies Into University Aerospace Engineering And Science Education*, Ad-Hoc Advisory Panel Meeting, NASA/LaRC, September 21-22, 2000.

growth in both the operating environment and the technical completeness of the technology. The environment becomes more representative of the real world as the technology moves from paper studies through laboratory setup and simulated environments to actual mission operations. You saw above that the operating environment for PBTs is people. A more representative environment of people implies that the community of people affected by the PBT becomes more representative of the general population of users as the PBT matures. The group of users expands from initial risk takers to more mainstream members of the community.

It's easy to measure the completeness of a hardware system. As the TRL goes up, the completeness of a system based on a hardware technology increases from basic scientific properties through breadboard components, integrated components, and prototypes, until the system achieves its final form. But the completeness of PBTs, applying knowledge using a new method or framework, is harder to measure. What is a "complete" method? Look at the organization of the knowledge. For a PBT, increasing completeness means the technology progresses from defined basic properties (research) through defined core practices, implementation mechanisms, and best practices (pilot studies), to a culmination in a final body of knowledge used across the entire community (mainstream users).

SEI used the usual method for adapting TRLs to a new type of technology. They changed the readiness level descriptions while leaving the basic definitions alone.

Modeling and Simulation Technologies

The Department of Energy, in a February 2000 presentation, suggested using TRLs to measure the maturity of modeling and simulation (M&S) technologies. They stated that TRLs for M&S systems are a function of three factors: correctness, usability, and relevance. They provided definitions for M&S TRLs 3 through 7 by slightly modifying the standard NASA hardware TRL definitions. Their presentation did not tell how to measure these newly defined TRLs.

CONCLUSION

The readiness level concept is flexible enough to be applied in many ways. This flexibility is both a blessing and a curse. When you misuse readiness levels, you contribute to the problem of readiness level proliferation. Properly designed new readiness levels, however, can help you monitor your program's performance in multiple dimensions.

I have shown you how the concept has been used to measure additional dimensions of hardware and software technologies. I called this use "other readiness levels." I also showed how the basic TRL concept can be modified to measure the maturity of other types of technologies that aren't strictly hardware or software technologies. I called these modifications to the standard TRLs "TRL corollaries," and I gave you four examples: bio-medical, learning system, practice-based, and modeling and simulation technologies.

© 2006 M. Engling

RISK AND VARIATION

CHAPTER OVERVIEW

I have introduced you to the Technology Readiness Level (TRL) scale as one measure of technology maturity. Now I want to investigate a different aspect of maturity, namely risk. I'll first show you what I mean by risk in a technology development program. I'll define risk and variation in technology development and show you how the two are related. I'll follow this by discussing how you can estimate risk from extremely limited data. I'll show you how technology development risk fits onto the whale chart. Finally, I'll discuss how the traditional view of risk can fall short of the experience of reality when unpredictable risk events are ignored.

In this chapter, I'll be showing you some statistical tools and measures of risk and variation. You should keep this warning from Thomas Pyzdek in mind, "Statistics are an expression of ignorance. They should only be used when ignorance is unavoidable, i.e., when knowledge is absent and

Did I Ever Tell You About the Whale? Or Measuring Technology Maturity, pp. 93–103
Copyright © 2008 by Information Age Publishing

unobtainable. Statistics are not knowledge. They are a calculation that permits action in the face of ignorance."

I recently read *The Black Swan: The Impact of the Highly Improbable* by Nassim Nicholas Taleb. This book caused me to re-think what I thought I knew about risk and variation. My new insights are summarized in the section, "Warning: Not All Risk Comes from Variation!" The earlier sections on risk and variation are not wrong. They are, however, incomplete. The traditional view of risk as the result of process variability is still legitimate in some settings, especially when the product or technology concerned is well understood and modeled. This view cannot predict the existence of an unpredictable disruptive threat to the status quo. Unfortunately, it is exactly this unpredictable threat that usually presents the greatest risk, i.e., the risk that has the most severe impact on research and development programs.

RISK AND VARIATION IN TECHNOLOGY DEVELOPMENT

Risk

Risk of program failure is an indicator of degree of program difficulty. Unfortunately, when you are trying to develop and mature a new technology, your ignorance is so complete that you'll have no experience to use as the basis for a statistical estimation of risk.

The traditional definition of risk involves two factors, probability of occurrence and consequence of occurrence. You can think of these as the chance that something will happen and how serious it will be if it does happen.

You can probably get a good idea of the second of these two factors. Brainstorming and other idea generating techniques can help you identify some potential foreseeable results of adverse happenings. Worst case analysis and failure mode and effects analysis can allow you to develop an appreciation for the consequences of the events you've identified. It's often easier to predict the effects of a risk event than it is to predict the event itself. Consider a weather example. You can predict roof damage as a possible consequence of bad weather without predicting any of the specific weather risk events that could cause the damage, such as hurricane winds, hail, ice dam, or tornado.

The first factor, calculating the probability of occurrence when you don't have experience to fall back on, is hard. You'll face problems in assigning probabilities to the identified risk events when you lack information and data that could be drawn from experience. We'll try to tackle these problems by approaching risk as an aspect of variation.

Variation

Risk and variation are so closely related that you can determine risk by estimating variation. Start with the assumption that everything you do will include some variability in its results when you perform it repeatedly. Let's look at a concrete example. Consider the process of driving to work every day. It's extremely unlikely that your driving time will be identical for every day of the year. Weather conditions will vary. Traffic density will be different, lighter on some days and heavier on others. Some days you'll make all the traffic lights while on other days every one will be red. An accident or road construction could cause major delays.

Risk and Variation

Now pretend that you're monitoring this drive-to-work process, and the driving time has to be less than a certain value or you get to work late, an unacceptable result. This maximum acceptable value, the upper limit of the acceptable range of possible values, is called the threshold. If the normally present process variation causes a measurement to fall beyond the threshold, you have experienced a failure. If you repeat the process many times, you can calculate the risk by finding the percentage of the results that fall outside the threshold as demonstrated in Figure 7.1.

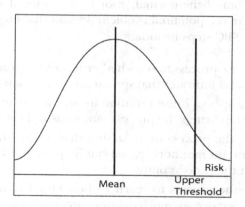

Figure 7.1. Risk as Probability of Failure

If you have established acceptable limits for a number of different requirements, your total risk can be defined as the probability that the result of the technology development effort will fail to meet one or more of the requirements. Probability theory lets you roll up individual risks to arrive at total risk. This means that all you have to do is calculate the risk associated with each individual requirement. Then you can find the overall project or program risk by rolling up the individual risk values into a single aggregate risk number. This works well when you have done a task enough times so that you can obtain the necessary statistics from the available data.

STATISTICAL PROCESS CONTROL

Remember that statistics is an expression of ignorance. Even when you have enough data to get statistical information, you'll still have to figure out what it all means. What are the numbers trying to tell you? What data points are meaningful? Is there a message hidden in the data, or is it just random noise? Statistical process control (SPC) is one method of pulling meaning from statistics.

Walter Shewhart established the statistical and philosophical basis for SPC in his 1931 book. The main idea of SPC is that a process can be controlled by controlling its variability or variation. Statistical analysis helps you identify variation and assign it to one of two categories, natural process variation or special cause variation. Natural process variation is the natural fluctuation and variability present in all processes. Sometimes it is called common cause variation or system variation. This contrasts with special cause variation which is variability in the process that resulted from a specific, out-of-the-normal, root cause. Special cause variation indicates that there is a potential problem present that ought to be fixed.

The four basic SPC steps include:

- Measuring the process to see whether it is out of control. An out-of-control process indicates that special cause variation probably exists.
- Eliminating special cause variation in the process to make it consistent. This step brings the process into statistical control.
- Monitoring the process to make sure that it stays in statistical control. By constant monitoring, you can fix problems before they seriously affect the process output.
- Improving the process to reach its best target value. In this step, you try to remove as much natural process variation as possible. This can be accomplished through process improvement techniques, such as investing in new equipment or reducing the number of process steps.

The control chart is a primary tool used by SPC. In fact, much of the existing SPC literature covers how to collect the necessary data and what control charts best portray the data. A common obstacle to successful use of SPC is getting bogged down with charts, forgetting that visual representation of data is a tool, not an end in itself. Control charts are useful because they provide you with a quick, visual means of identifying problems in your process.

In the SPC chart shown in Figure 7.2, I plotted the number of defects discovered during inspections of a series of quality control samples. According to SPC theory, all data points that fall between the upper and lower control limits indicate that the process is showing only natural process variation. In my sample chart, most of the points fall between the limits; however, there are two points beyond the control limits. These two out-of-control points indicate measurements that probably include some special cause of variation.

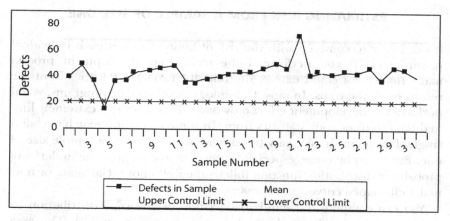

Figure 7.2. Sample Statistical Process Control (SPC) Chart.

Look at the point that's above the upper control limit. There were more defects present in this sample than you would expect from the normal variation in the process. Something went wrong here. Your job is to find out what went wrong, identify the root cause, and fix it so you don't have this problem again.

Take a look at the point that fell below the lower control limit. This data point is also out-of-control, even though it's good to be below the expected number of defects. What should you do with this data point? You should try to find out why this sample was so exceptionally good. If you could capture the reason for this outstanding result and make it part of the process, you might make the overall process better and improve quality for everyone.

How about all the other points? The message of the in-control points is simple, "If it ain't broke, don't fix it." Tinkering with an in-control process will almost always make it worse. Trying to find the root cause of variation for data that's fluctuating between the control limits is a futile exercise in chasing the data. Normal process variation has no root cause that you can identify and fix. If you can't live with the number of defects you get from a process that's in-control, you'll have to do some process engineering to improve the process so that it is capable of meeting your needs. Trying to fix the problem without improving the process won't work.

SPC is used most often with manufacturing processes. SPC relies on sample data, so you need to have enough data available to analyze statistically. The lack of data in the laboratory environment makes SPC of limited value early in the technology life cycle.

ESTIMATING RISK FROM A SAMPLE OF SIZE ONE

It's not easy to come up with the risk of failing to meet each individual requirement. If you performed the technology development process many times, you could extract the relevant variation statistics from the results of all the trials. In most technology development programs, you'll perform the development effort only once. This gives you extremely limited data to draw your statistics from. In some cases, you may have but a single data point. It isn't easy to estimate variation from a sample size of one. But it can be done, especially if you can assume that the underlying probability distribution function follows the well-known Gaussian, or normal, bell-shaped curve.

You can compute the standard deviation of a normal distribution by estimating the height of the bell-shaped curve (Figure 7.3). The area under the curve represents a probability. This means that the total area must always be equal to one. As the height of the curve increases, the curve must become more peaked to keep the area constant. Conversely, as the height of the curve decreases, the curve must flatten out. A peaked curve has less variation than a flat curve does. The numerical measure of variation around the expected value, or mean, is the standard deviation.

A curve with a large standard deviation is flatter than a curve that has a smaller standard deviation. Similarly, a curve with a small standard deviation is more peaked than one with a larger standard deviation.

If you can determine the height of the curve, you can use the equation for the Gaussian distribution to find the standard deviation. One way to estimate the height is to define it as how believable the expected value appears to be. The greater your believability in the expected value, the higher the curve will be. A high curve implies a low standard deviation, as you have just seen.

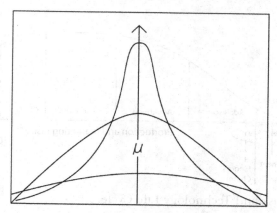

Figure 7.3. Flatness vs. Height of Curve.

Let's see what we'll know at this point in our requirement analysis. You should know the minimum or maximum acceptable value (threshold) for this particular requirement. You'll have some idea of the expected value, and you'll know how confident you are that this value is correct. From this believability value, you can compute an estimate for the standard deviation. Knowing the standard deviation and the expected value tells you everything you need to know about the normal curve. You can easily compute risk by calculating the percent of the curve that falls outside the threshold.

It's really tempting to say, "Details are left as an exercise for the reader;" however, I have included Appendix C to provide more information for you if you are interested.

DID I EVER TELL YOU ABOUT THE WHALE? RISK AND THE TECHNOLOGY LIFE CYCLE

The whale chart shows you where development risk falls in the life cycle. It should come as no surprise that this sort of risk occurs early. Other types of risk, such as production risk, marketing risk and risk of obsolescence exist at later stages of the technology life cycle.

As you can see from Figure 7.4 on the next page, development risk occurs in advancing the technology's readiness level from TRL 1 through TRL 6. You could argue that this type of technical risk continues until the technology has achieved TRL 9. The dashed line in the whale chart represents this option. I like to end technology development risk at TRL 6,

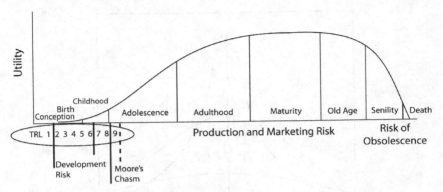

Figure 7.4.   Risk and Technology Life Cycle.

because in going from TRL 6 to TRL 9, the risk is in product development as opposed to pure technology development in the laboratory.

The figure also shows, as you saw in Chapter 3, that the marketing risk Geoffrey Moore calls the chasm occurs toward the end of technology development. Once a technology has reached maturity, risks consist mainly of failure to achieve or maintain market share. Production risks, such as material shortages and labor problems, can also occur here. Finally, and I mean this in its absolute sense, risk of obsolescence happens very late in the technology life cycle. I didn't cover any of these other risk types. My discussion was limited to development risk.

WARNING: NOT ALL RISK COMES FROM VARIATION!

Earlier in this chapter, I told you, "The traditional definition of risk involves two factors, probability of occurrence and consequence of occurrence." This view of risk includes some unstated assumptions which can be derived from the idea that risk, as I argued earlier, is so closely related to variation that you can determine risk by estimating variation. In this section, I'll discuss these usually unspoken assumptions, which can have both positive and negative consequences. On the plus side, the assumptions allow you to treat risk quantitatively, calculating or estimating the probability that a given program parameter will fail to meet your customer's needs. Rolling up all of the individual risks gives you an estimate of overall program risk. This overall program risk estimate can be used to compare two or more research and development programs. There are commercial software programs[1] available that automate this risk estimation process.

On the down side, the idea that risk springs entirely from variation can hide some serious program risks. The idea that all risk comes from variation implies that the parameters of interest are well known, the processes for producing the parameters are understood, and risk can be calculated by computing the portion of the probability distribution falling outside the threshold. These three factors are not always present. Sometimes there is no way to predict a specific risk event or even compute the probability that it will occur. Unpredictable events often have great impact when they do happen, yet these high-impact, highly unlikely events are not covered by the traditional Gaussian view of risk.

Assumptions in Traditional Risk Model

In the traditional risk model, risk is defined as the probability that the measured value of a parameter of interest will fall outside the range of values that satisfy your customer's requirements. Calculating risk then becomes an exercise in computing probability. Recall that risk was defined as involving two factors: probability of occurrence and consequence of occurrence. Normal process variation affects the probability of occurrence by allowing parameter values to fluctuate about an average value, the mean. Wide departures from the mean resulting from normal variation could cause a parameter value to fall beyond the acceptable threshold. This is a risk event.

Statistical probability theory will give you the tools needed to perform these computations if you can assume that the following statements apply to your situation:

The underlying process model for the risk parameters of greatest interest is Gaussian.

Normal process variation will cause fluctuating values around the mean for each parameter of interest.

Risk events can be statistically predicted from data. When data is limited or unavailable, risk events can be estimated as discussed in the section "Estimating Risk from a Sample of Size One."

Traditional Risk Mitigation

When these assumptions hold, traditional risk mitigation techniques work well. The primary traditional risk mitigation method is to reduce the probability of occurrence by controlling process variation. When you control the amount of variation possible in your process, you lower the standard deviation, and the parameter values cluster more closely around the expected value. The tails of the distribution curve become so small that they can be ignored; your probability of falling into this area of the curve is really small.

Once your process variation is controlled, it is extremely unlikely that you will experience a risk event. Your risk probability is low enough to no longer be a concern. Monitoring the process on a continuous basis, possibly using the SPC technique discussed earlier, will keep your process under control. Voila! Your risk management and mitigation problem is solved!

Assumptions in Non-Gaussian Model

Unfortunately, the real world isn't as predictable as the traditional Gaussian model would have you believe. In real life, the happenings that have the most impact are almost always totally unexpected. This also applies to risk events. Those that have the greatest impact on our project or program are the risks that COULD NOT BE PREDICTED. The emphasis is intentional. High impact risk events are, by nature, totally unpredictable. This doesn't just mean that you failed to predict these risks; it means that there was no way that anyone could possibly have predicted them.

High impact risk events are unpredictable because they are also extremely unlikely. They are not the result of random variation about some process mean value. These risk events do not result from variation in a known process. They come from outside your experience base. The underlying process model for these risk events of greatest impact and interest is non-Gaussian, and possibly non-linear.

The key assumption is simply this: the highest impact risks events cannot be predicted.

Non-Gaussian Risk Mitigation

Since this type of risk cannot be predicted and is not caused by variation, you won't be able to mitigate it by traditional risk management methods. Controlling process variation as a means of reducing the probability of occurrence won't work. Remember that these high impact risk events are unpredictable because they are so unlikely. Since the probability of occurrence is already low, you can't lower it any more by controlling variability. You're already way out in the highly unlikely tail of the probability distribution curve!

Recall once again that risk involves both the probability and severity of the occurrence of a risk event. You see at once that, since you can't lower the probability of occurrence, your only available risk mitigation strategy is to reduce the severity of the consequence of the event. But how can you reduce the severity of the consequences when you can't predict what will happen? Way back at the start of this chapter, I told you that it's often easier to predict the adverse results of a risk event than it is to predict the event itself. Look at a hypothetical technology development effort. Pretend you're working on developing a laser application for a high power laser. There might be a lot of reasons why your application could fail. One type of failure might be that the laser doesn't put out enough power. Note that you have not specified the cause of this failure. You aren't going to be able to prevent it from happening by controlling manufacturing process variation, since you don't have a specific cause. But you could mitigate the risk by developing two or three different types of lasers, picking the best one later.

Risk mitigation in the non-traditional, non-Gaussian case is about reducing the impact of an unknown, unpredictable risk event.

CONCLUSION

This chapter introduced you to the idea of risk in a technology development program. You also discovered that estimating risk from a limited sample is not easy, since there isn't enough data to support usual methods of statistical analysis. I showed you one useful method of estimating risk from a sample size of one using your degree of belief in the correctness of the sample value. I put development risk and some later risks on the whale chart. Finally, I talked about some limitations of this variation and risk duality by briefly introducing the idea of unpredictable, high-impact risk events. In the next chapter, you'll see how risk estimation can help you determine how difficult your remaining development work might be.

NOTE

1. *Dynamic Insight* marketed by James Gregory Associates, Inc., is one product known to the author.

© 2006 M. Engling

© 2006

HOW HARD CAN IT BE?

"It's Tough to Make Predictions, Especially About the Future."

Yogi Berra

CHAPTER OVERVIEW

I have introduced you to the Technology Readiness Level (TRL) scale as one measure of maturity for a technology in early development. TRLs give you a way to define both where you are and where you want to be. Now I want to help you answer the question, "How hard will it be to go from where I am to where I want to be?"

RESEARCH AND DEVELOPMENT DEGREE OF DIFFICULTY

NASA's John Mankins introduced the research and development (R&D) degree of difficulty (R&D^3) as an additional measure to complement TRLs. He proposed that R&D^3 would measure how much difficulty you could expect in attempting to mature a given technology. His 1998 white paper recommended five distinct levels of difficulty. Each of the R&D^3 levels can be described by the probability that your R&D effort would be successful given that you put in a "normal" amount of work. These probabilities of success run from almost certain failure at R&D^3 Level V to almost certain success at R&D^3 Level I. The five levels are shown in Figure 8.1.

Mankins assigned a probability of success to each of the R&D^3 levels as follows:

R&D^3 Level I — 99% or almost certain success.
R&D^3 Level II — 90%.
R&D^3 Level III — 80%.
R&D^3 Level IV — 50%.
R&D^3 Level V — 10% to 20% or almost certain failure.

The Mankins R&D^3 measure was a step toward defining R&D program risk by attempting to predict the difficulty of advancing a technology's maturity. Mankins defined the five degrees of difficulty and gave examples of each, but his white paper gives no information on how to apply this measure. Other researchers approached risk definition from a different direction. I'll now introduce you to a measure based on amount of effort remaining.

TECHNICAL PERFORMANCE MEASURE RISK INDEX

Paul Garvey and Chien-Ching Cho said that the risk remaining in a technology development program at a given time is the same as any unmet performance requirements at that time. The performance requirements come from the program's technology performance measures (TPMs), hence the name of the index. For every TPM, you will have to demonstrate that you have achieved at least a specific value, known as the threshold, to claim acceptable performance for that measure. In their paper, Garvey and Cho showed how you may combine individual TPMs to measure and monitor the overall performance risk of a system.

Garvey and Cho recognized that your program's individual TPMs can fall into one of two types:

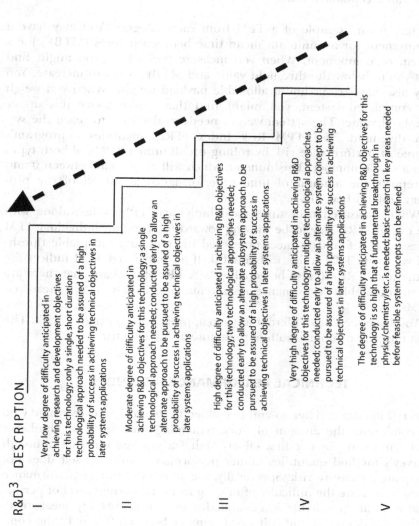

R&D³   DESCRIPTION

I   Very low degree of difficulty anticipated in achieving research and development objectives for this technology; only a single, short duration technological approach needed to be assured of a high probability of success in achieving technical objectives in later systems applications

II   Moderate degree of difficulty anticipated in achieving R&D objectives for this technology; a single technological approach needed; conducted early to allow an alternate approach to be pursued to be assured of a high probability of success in achieving technical objectives in later systems applications

III   High degree of difficulty anticipated in achieving R&D objectives for this technology; two technological approaches needed; conducted early to allow an alternate subsystem approach to be pursued to be assured of a high probability of success in achieving technical objectives in later systems applications

IV   Very high degree of difficulty anticipated in achieving R&D objectives for this technology; multiple technological approaches needed; conducted early to allow an alternate system concept to be pursued to be assured of a high probability of success in achieving technical objectives in later systems applications

V   The degree of difficulty anticipated in achieving R&D objectives for this technology is so high that a fundamental breakthrough in physics/chemistry/etc. is needed; basic research in key areas needed before feasible system concepts can be refined

Figure 8.1.   R&D Degrees of Difficulty.

- More is better. The current value of this TPM must increase to reach an acceptable level.
- Less is better. The current value of this TPM must decrease to reach an acceptable level.

Here is an example of a TPM from each category. You may have a requirement for a minimum mean time between failures (MTBF) for a system or component. When you measure the MTBF, you might find that you're below the threshold value and MTBF needs to increase. You may also have a maximum allowable payload weight. When you weigh the completed system, you might find that you're above this upper threshold value. The system weight needs to decrease to reach the system threshold. The TPM Risk Index (TRI) quantifies a program's degree of performance risk by rolling up all unmet TPMs of both types into a single number. This number, which will always fall between 0 and 1, reports the amount of unmet performance across all of the program's established TPMs.

When measured over time, each individual TPM index, along with the overall TRI, should be moving toward 0. When an individual TPM is tested and found to meet or exceed its minimum acceptable threshold value, its TPM index will equal 0. If all individual TPM indices are 0, the total TRI will also be equal to 0. A TRI of 0 means that there are no unacceptable performance risks remaining in any of the measured TPMs.

You have seen two different ways to approach R&D program risk. The next measure combines both of them into a single metric.

TECHNICAL PERFORMANCE RISK INDEX

Sherry Mahafza and her co-authors added to the work of Garvey and Cho by combining the amount of work remaining measure with Mankins's R&D^3 measure. In her first effort, Mahafza pointed out that although Garvey's method quantifies unmet performance, this in itself does not adequately measure risk. Specifically, the amount of unmet performance doesn't measure the difficulty of moving from the current level of performance to an acceptable level. Mahafza used a probability measure to define the degree of difficulty as a number between 0 and 1. She combined this with the amount of effort remaining to come up with the technical performance risk index (TPRI).

In her later work, Mahafza modified the earlier TPRI to better model its behavior near the extreme values, e.g., certain success (degree of difficulty = 0) and certain failure (degree of difficulty = 1). Following Garvey's lead, she showed that when there is almost guaranteed success, the program's technical risk is equal to the unmet performance requirement. As the degree of difficulty increases, so does the risk of R&D program failure. When the difficulty of improvement is very high, no matter how close your achieved performance is to the required performance, you still have high risk of failure.

I'll give you a made-up example to show you what all of this means. Suppose you're required to demonstrate a 10 watt output from your R&D device, and so far you've managed to get 9.5 watts. Garvey and Cho would say that you have demonstrated 95% of the threshold, so the remaining risk is only 5%, or 0.05. Their argument would go something like this: Since you have demonstrated 95% of the requirement, it's only the last 5% that's at risk. Mahafza says that this is true only if you're almost guaranteed to get to 10 watts if you continue working. If there's any possibility of not reaching the threshold, then amount of unmet performance won't adequately measure risk. Maybe the 95% you have achieved was the easy part, but the little bit remaining can't be done without a breakthrough in our fundamental knowledge of the physical universe. In this case, the 5% remaining effort would not be an accurate measure of program risk. That's why she added expected difficulty of improvement to the measure.

ADVANCEMENT DEGREE OF DIFFICULTY

The final metric you'll meet is really more than a measure; it's a complete method for evaluating R&D program risk. James Bilbro of NASA has expanded Mankins' $R\&D^3$ idea into what he calls the Advancement Degree of Difficulty (AD^2). AD^2 is a method of dealing with other aspects beyond TRL. It describes what you are required to do to move a technology from one TRL to another. Since it starts with a known TRL, AD^2 is the second stage of a two-stage readiness assessment process. The first stage is determining the current and target TRLs. Once you have identified the key technologies and assigned their respective TRLs, you must determine what is required to advance them to the level necessary for the success of your R&D program. Bilbro says that this AD^2 assessment is one of the most challenging aspects of technology development for two reasons: all technologies are not the same, and the assessment requires the art of prediction. Making predictions is tough, as you saw at the start of this chapter.

The big improvement of AD^2 over $R\&D^3$ lies in the way that AD^2 provides a method of application. Bilbro gives detailed guidance on how to apply this measure. He starts by assembling a team of experts. To emphasize the need for getting the right people on the assessment team, Bilbro said, "It is extremely important that you go wherever you need to in order to get the expertise needed!" The necessary experts will include scientists, technologists and engineers from all required disciplines as well as program managers and outside consultants if required.

Once the team of people possessing the appropriate expertise has been assembled, "all" that you need to do for a successful AD^2 is determine what steps are necessary in the technology development, decide what tests you need to perform to certify the development, and make an informed assessment (guess) as to the degree of difficulty you expect to encounter in advancing the technology to meet the R&D goals. AD^2 also considers the type of demonstration unit the program will be required to produce. Technology development includes the fabrication of any laboratory mock-ups, breadboards, brass boards, developmental models, or prototypes you will need either for research or for test and evaluation. AD^2 takes the five levels of R&D[3] and applies them to four areas of concern:

- Design and Analysis.
- Manufacturing.
- Test and Evaluation.
- Operability.

I'll describe each of these four areas in turn.

Design and Analysis

In this area you are concerned with whether or not the necessary design methods, models, analytical tools, and data bases are available for the design of the R&D hardware or software. You will also try to discover whether you have everything you'll need for the analysis of any experimental results.

Manufacturing

The manufacturing area is concerned with two separate things. You want to know whether you can manufacture the development units needed to accomplish the goals of the R&D program. You also want to know whether you will be able to manufacture the final product that the technology development is meant to support. In each of these separate topics, you want to know whether the necessary tools, materials, and manufacturing and quality processes are available.

Test and Evaluation

Test and evaluation (T&E) as used in this context deals with the formal process that certifies the results of the R&D effort. It does not include the on-going testing and evaluating of experiments that is incidental to any laboratory investigation. Here you're concerned with the tests and evaluations that are performed at the end of the development effort. Usually, this T&E is separated into two types. Developmental T&E (DT&E) answers the question, "Does the finished developmental item meet the requirements and specifications established for the R&D program?" (Did you build the thing right?) Operational T&E (OT&E) tests the final product under real world conditions to see whether or not it can fulfill the need you were trying to fill. (Did you build the right thing?) In performing an AD^2 assessment of T&E, you want to make sure that you have everything you'll need to perform these tests at the end of the R&D effort. "Everything" includes facilities, equipment (hardware and software), trained personnel, and any other special requirements.

Operability

Throughout the development of the design, manufacturing and testing processes, you must take operability into account. Operability is more than just how easy it is to operate the product. Operability also covers all of the so-called "-ilities." Here is where you will formally document any concerns about reliability, maintainability, supportability, availability, verifiability, testability, and reproducibility. You'll also consider life-cycle costs. You could make a case for including ease of manufacture here under the producibility banner, but that topic is already included under manufacturing, which is where it seems to belong.

AD^2 Wrap-Up

The AD^2 assessment gives you a lot of information about an R&D technology development program. Bilbro captures the information in an AD^2 Assessment Matrix. This format makes it easy to see at a glance which portions of the development effort require the most management attention. Using the matrix, it is possible for you to roll up individual AD^2 ratings to come up with a single overall program or technology degree of difficulty.

A comprehensive AD^2 assessment gives you much of the information needed to develop a technology roadmap. Once the assessment is complete, it's easy to identify critical technologies. Because you know what remains to be done and how difficult the remaining tasks are, you should be able to create realistic schedules. Since you know what tools, processes, equipment, and facilities capabilities you lack, you should be able to develop reasonable schedules, cost plans, and budgets.

The AD^2 process is more fully described in Appendix D, written by Jim Bilbro. You will also find a copy of a software tool that automates the AD^2 process with the companion software that accompanies this book.

CONCLUSION

You have learned about several methods available to you when you try to answer the question, "How hard can it be to mature this technology?" You have also seen that several researchers are continuing to work in this area, so the best method available to you today might not be the best in the future. Jim Bilbro's AD^2 is currently the best predictive method for determining the difficulty of further advancement in a technology's maturity because it is the method that is most comprehensive and also the method with the most developed application procedure. The AD^2 assessment can be the key to sound program management in the uncertain world of technology development.

© 2006 M. Engling

TECHNOLOGY
MATURITY ASSESSMENT

CHAPTER OVERVIEW

In this chapter, you'll discover several ideas about how best to perform a technology maturity assessment (TMA). Recall that in Chapter 4 you learned that the technology readiness level (TRL) is a measure of a technology's maturity at a given time. First, I'll show you some ideas that deal with how inclusive the assessment should be. Then you'll see how a stage gate process can be used to perform maturity assessments. I'll tell you what both the Department of Defense (DoD) and NASA currently require in their TMAs before giving my recommendations for a complete TMA.

Did I Ever Tell You About the Whale? Or Measuring Technology Maturity, pp. 113–118

WORK BREAKDOWN STRUCTURE

An all-inclusive TMA might include every single item in a system's work breakdown structure or WBS. The WBS is a hierarchical breakdown of the hardware and software products of the program or project. It is often displayed in a table where each WBS element is given an identifying number that specifies its position in the project hierarchy. For example, if WBS element number 1 is an air vehicle, element 1.1 might be the fuselage, 1.2 the wing, and 1.3 the empennage. Carrying the WBS further, element 1.2.1 could be the main spar, etc.

NASA argues that a complete TMA ought to address every element in the WBS. The argument runs something like this: until you've looked at the maturity of every WBS element in its proposed operating environment and configuration, you can't be certain of your overall maturity. By addressing every single item in the WBS, you're sure not to overlook anything. As you will find in Appendix D, even though a heritage or legacy item has been previously used in actual mission operations (TRL 9), NASA wants you to downgrade that item to TRL 5 until and unless you can justify a higher TRL value based on its present configuration and use. Remember, if you're using an existing technology item in a new configuration or in an environment in which it has never been tried, the item cannot be considered to be mature.

CRITICAL TECHNOLOGY ELEMENT

What if you don't want to take the time to perform a TMA on each and every item on the WBS? Maybe the system is so huge and complex that there are just too many WBS elements to address individually, and many of them might use nothing but well-known technologies. In such a case, you may want to perform a TMA on just a subset of the WBS items. Fortunately, DoD has defined Critical Technology Elements (CTEs) as the particular subset of technology elements that you need to perform a TMA on.

Which technology elements are critical? I can almost hear you asking yourself that question. Once again, it's DoD to the rescue. The TRA Deskbook definition says, "A technology element is 'critical' if the system being acquired depends on this technology element to meet operational requirements (with acceptable development, cost, and schedule and with acceptable production and operation costs) *and* if the technology element or its application is either new or novel. Said another way, an element that is new or novel or is being used in a new or novel way is critical if it is necessary to achieve the successful development of a system, its acquisition, or its operational utility."

Note the italicized "and" in the CTE definition. It's important. A CTE must meet both parts of the definition. The first half of the definition says that for an element to be defined as critical, the system must depend on it to meet its operational requirements. Notice that "operational requirements" include "acceptable" development cost and schedule as well as "acceptable" production and operation costs across the system's planned life cycle. The other half of the definition requires the element "or its application" to be new or novel. A brand new technology element meets this requirement. So does an existing element that is being used in a new configuration or in an environment in which it has never been tried. Just as NASA says in Appendix D, DoD says that an existing technology that's being used in a new way cannot be considered to be mature; you'll have to do a TRA on such an item.

This brings us to the next point. A TRA must cover all of the CTEs (but only the CTEs) included in the system under development. CTEs may be related to any dimension of the technology development effort; hardware, software, manufacturing, or life cycle (supportability and disposal). CTEs may also be identified at the subsystem or component level as well as at the overall system level. In other words, any WBS element could potentially include a CTE.

STAGE GATE

A stage gate process divides an overall development effort into stages, or phases, separated by decision points, or gates. As you complete each development stage, you must demonstrate that you have met the criteria necessary to successfully pass the decision gate before you can begin work on the next program phase. Every gate presents you with a multiple choice decision. You can proceed to the next stage, remain in the current stage, revisit an earlier stage, or kill the program outright. For a technology development program, a stage gate process can be an effective maturity assessment process. In fact, the DoD Acquisition Milestones were created to act as decision gates for major defense acquisition programs.

One problem with the DoD Acquisition Milestones is that they lack granularity. The program phases take too long, so too much time passes between gate reviews. A large acquisition program could be on track at Milestone B, but, long before it comes due for a Milestone C decision, things could go bad. More detail between gates might solve this difficulty, letting you catch problems while they're still small.

TECHNOLOGY PERFORMANCE MATURITY MODEL

The U.S. Army's Space and Missile Defense Command has developed a stage gate process that puts a decision point at every TRL change. Their system results in more frequent gate reviews because you can't move from one readiness level to the next unless you not only demonstrate the technical performance necessary, but also meet all of the other programmatic and documentation criteria appropriate at each level.

The DoD TRA Deskbook is also taking DoD down this path. As you saw in Chapter 4, at each TRL the deskbook defines supporting information that must be provided to justify your claim that your technology has achieved this level of maturity. In short, it looks like both the Army and DoD are moving toward a stage gate maturity measurement process based on TRLs.

DEPARTMENT OF
DEFENSE TECHNOLOGY READINESS ASSESSMENT

DoD has published a reference, the DoD Technology Readiness Assessment (TRA) Deskbook, that provides detailed instructions as to how and when to perform these assessments. The deskbook is available on the world-wide web. In a nutshell, DoD requires you to do a TRA at or before acquisition milestones B and C. There are minimum maturity requirements that must be met at these two milestones. At Milestone B, the system or subsystem model or prototype must be demonstrated in a relevant environment. This, as you learned in Chapter 4, is the same as TRL 6. At Milestone C, the maturity requirement (equivalent to TRL 8) is for a full system prototype demonstration in an operational environment.

The TRA is performed only on the Critical Technology Elements (CTEs) as defined earlier in this chapter, and consists of a TRL determination for each of them. In fact, DoD defines the TRA as, "A systematic, metrics-based process that assesses the maturity of CTEs." It uses TRLs as the maturity metric.

It's important to realize that in DoD a TRA is not a risk assessment. Since the metric is the TRL, a TRA only can tell you where you are on the maturity scale. It cannot tell you how much difficulty or risk your technology faces in advancing further if it's not already as mature as it needs to be.

Since a DoD TRA does not cover every item in the WBS, it can't be the same as a design review. The TRA covers only the CTEs, and specifically does not address integration. It's because of these limitations that I do not consider the DoD TRA as representative of a complete TMA.

NASA TECHNOLOGY READINESS ASSESSMENT

A detailed discussion of the NASA TRA process is included in Appendix D, so I won't cover the details here. You ought to know that NASA has a two-part TRA process. For every item in the WBS, you are required to perform a readiness assessment based on the TRL. For each WBS element that is not at the necessary level of maturity, you also perform an Advancement Degree of Difficulty (AD^2) assessment.

The idea is that the two-phase approach covers the entire system because it is WBS based. It also estimates risk through the AD^2 measurement process.

A COMPLETE TECHNOLOGY MATURITY ASSESSMENT

In my opinion, a complete technology maturity assessment (TMA) has three parts. First, you need to identify all of the CTEs as discussed above. CTEs will include all newly developed (or developing) technologies that are essential for the operation or acquisition of the entire system. This includes existing technology elements (think of off-the shelf technologies) that are being used in new or different ways. The WBS is a great data organizer for you to use to make sure you don't overlook any critical elements as you define the system CTEs.

Once you're sure that you have identified all of the CTEs, you'll need to perform a TRA to determine the TRL of each one. As mentioned in Chapter 4, the AFRL TRL Calculator is a tool that will help in this task. Make sure that you perform a TRA on every one of the CTEs. This is especially important if you're using an old technology in a new way. Remember, NASA says that you ought to downgrade a TRL 9 technology all the way back to TRL 5 if it's being used in a new way or in an untested environment.

The third and final step in a complete TMA is an AD^2 assessment which must be performed on every CTE that has not yet reached its target TRL. The AD^2 assessment will help you determine which CTEs carry the greatest risk of failing to advance to the maturity level required. You should be aware, however, that even when you include an AD^2 assessment in your TMA, you still have not done a formal risk assessment. While you have an indication of risk areas, you have not identified specific risks, nor have you initiated a risk tracking process and developed a risk mitigation plan for each risk. All of these would be needed to convert your TMA to a total maturity and risk evaluation.

CONCLUSION

This concludes my discussion of TMAs. I hope to leave you with the idea that, in my opinion, neither NASA nor DoD policy requires a complete TMA. Even if a full TMA were accomplished on a specific technology development program, other management assessments, particularly in the area of risk, would still be necessary for you to have a complete picture of the program's overall status.

© 2006 M. Engling

© 2006

Epilogue

DID I EVER TELL YOU ABOUT THE WHALE?

I've told you quite a bit about the whale; however, as you can see in Figure 1, you have been reading about measuring the maturity of technologies only during their initial development. This entire book has been about the tail of the whale, and that's no fluke. Most of the problems you'll encounter in working with technology maturity will happen early in the life cycle. It's while you're developing a new product or system that you are likely to encounter difficulties from using immature technologies, as the Government Accountability Office continues to point out. That's why almost all of the research on technology maturity that's been performed to date has concentrated on the whale's tail. Even with all of this work, there is a need to do more research on the tail of the whale. I have tried to show where there are gaps in our current understanding, and I also showed you several areas where there are opportunities to develop standard methods of measuring maturity and risk.

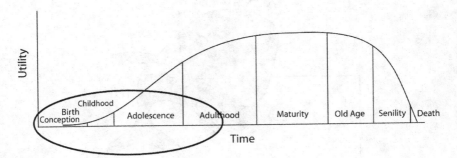

Figure 1.   Stages of Technology Maturity—The Whale Chart.

There is a serious need to do research work that would extend technology maturity measurement beyond the tail. We all would benefit from the availability of tools that could help us not only react to product or technology obsolescence, but also predict and anticipate problems in the body and at the head of the whale. This is definitely an area that is ready for intensive effort; however, the bulk of the whale remains beyond the scope of the present book.

© 2006 M. Engling

© 2006

Appendix A

Acronyms and Abbreviations

AD^2	Advancement degree of difficulty, Bilbro
AFRL	Air Force Research Laboratory
BOGSAT	Bunch of guys sitting around a table
CD	Compact disk
CECOM	Communications-Electronics Command, U.S. Army
CERDEC	Research, Development and Engineering Center, CECOM
CMM	Capability Maturity Model, SEI
CMMI	Capability Maturity Model for Integration, SEI
COTS	Commercial off-the-shelf
DoD	Department of Defense
DT&E	Developmental test and evaluation
DTC	Design to cost
EMRL	Engineering and Manufacturing Readiness Level
FRP	Full Rate Production
GAO	Government Accountability Office, formerly General Accounting Office
GOTS	Government off-the-shelf
H/W	Hardware
I/O	Input and output

Table continues on next page.

Table Continued

IAP	Independent advisory panel, NASA
IPPD	Integrated Product and Process Development
IPT	Integrated Product Team
IRL	Integration readiness level
ITWG	Information Technology Working Group, DoD, 2004
KDP	Key decision point, NASA
KP	Knowledge point
LRIP	Low Rate Initial Production
LRL	Logistics readiness level
M&S	Modeling and simulation
MANTECH	Manufacturing technology
MDA	Missile Defense Agency
MOD	Ministry of Defense, UK
MRL	Manufacturing readiness level
MTBF	Mean time between failures
MTTR	Mean time to repair
NASA	National Aeronautics and Space Administration
NASA/LaRC	NASA/Langley Research Center, Hampton VA
OSD	Office of the Secretary of Defense
OSD/ATL	Deputy Under Secretary of Defense for Acquisition, Technology and Logistics
OT&E	Operational test and evaluation
PBT	Practice-based technology, SEI
PDR	Preliminary design review
PRL	Programmatic Readiness Levels; formerly PRT
PRT	Program Readiness for Transition; now PRL
R&D	Research and development
$R\&D^3$	Research and development degree of difficulty, Mankins
RFI	Request for information
S&T	Science and technology
SEI	Software Engineering Institute, Carnegie-Mellon University
SEMP	Systems Engineering Master Plan
SPC	Statistical process control
SRL	Supportability readiness level; sustainment readiness level
SWRL	Software Readiness Level
T&E	Test and Evaluation

TA	Technology assessment, NASA
TAR	Technology assessment report, NASA
TEMP	Test and Evaluation Master Plan
TMA	Technology maturity assessment, NASA
TPM	Technology / technical performance measure
TPRI	Technology performance risk index, Mahafza
TRA	Technology readiness assessment
TRAR	Technology readiness assessment report, NASA
TRI	TPM risk index, Garvey and Cho
TRL	Technology readiness level
UK	United Kingdom
VV&A	Verification, validation and accreditation
WBS	Work breakdown structure

© 2006 M. Engling

RECOMMENDED READING
LIST BY CHAPTER

CHAPTER 1

_____, "Better Management of Technology Development Can Improve Weapon System Outcomes," GAO/NSIAD-99-162, July 1999.

_____, "Levels of Information Systems Interoperability (LISI)," C4ISR Architecture Working Group, Department of Defense, March 30, 1998.

Bilbro, James W., "Technology Readiness and Roadmapping," presentation, NASA, April 25, 2001.

Bilbro, James W., "The Impact of Technology Maturity on Program/Project Success – And How to Mitigate It," presentation at the Aerospace Corp. Technology Readiness and Development Seminar, NASA, April 28, 2005.

Davidow, William H., _Marketing High Technology, an Insider's View_, New York, 1986.

Garvey, Paul R. and Chien-Ching Cho, "An Index to Measure a System's Performance Risk," _Acquisition Review Quarterly_, Spring 2003, pp. 188–199.

Mankins, John C., "Research and Development Degree of Difficulty (R&D3)," NASA White Paper, 1998.

CHAPTER 2

_____, "Best Practices – Successful Application to Weapon Acquisitions Requires Changes in DOD's Environment," GAO/NSIAD-98-56, February 1998
_____, "Operation of the Defense Acquisition System," DOD Instruction 5000.2, 12 May 2003.
Carr, Nicholas G., "IT Doesn't Matter," published in *Harvard Business Review*, May 2003.
Institute of Electrical and Electronics Engineers Standards Board, "IEEE Standard for Application and Management of the Systems Engineering Process," IEEE-Std 1220-1998, 8 December 1998.

CHAPTER 3

Beshears, Fred M., "The Technology Adoption Life-Cycle," University of California Berkeley, 1999.
Doyle, Peter and Susan Bridgewater, eds., *Innovation in Marketing*, Oxford, 1998.
Electronics Industry Association (EIA) Engineering Department, "EIA Standard, Product Life Cycle Data Model," EIA-724, September 1997
Komninos, Ioannis, "Product Life Cycle Management," Thessaloniki 2002.
Moore, Geoffrey A. *Crossing the Chasm*. Harper Business, 1991.
Onkvisit, Sak and John J. Shaw, *Product Life Cycles and Product Management*, Westport, 1989.
Steinhardt, Gabriel, "Extending Product Life Cycle Stages,"2003.

CHAPTER 4

_____, "AMS Guidance on Technology Readiness Levels (TRLs)," UK MOD, February 4, 2002.
_____, "Better Management of Technology Development Can Improve Weapon System Outcomes," GAO/NSIAD-99-162, July 1999.
_____, "Defense Acquisition Guidebook," Department of Defense, November 2004.
_____, *Technology Readiness Assessment (TRA) Deskbook*, Department of Defense, May 2005.
Bilbro, James W., "Technology Readiness and Roadmapping," presentation, NASA, April 25, 2001.
Buehler, Martin, PhD, "New Millennium Program Description of Technology Readiness Levels", NASA, April 2002.

Graettinger, Caroline P., PhD, Suzanne Garcia, Jeannine Siviy, Robert J. Schenk, and Peter J. Van Syckle, *Using the "Technology Readiness Levels" Scale to Support Technology Management in the DoD's ATD/STO Environments*, CMU/SEI-2002-SR-027, Software Engineering Institute, August 2002.

Mankins, John C., "Technology Readiness Levels," a white paper, NASA, April 6, 1995.

Nolte, William L., "AFRL Hardware and Software Transition Readiness Level Calculator, Version 2.2," 2004.

Sadin, Stanley T.; Povinelli, Frederick P.; Rosen, Robert; "NASA Technology Push Towards Future Space Mission Systems," Acta Astronautica, V 20, 1989, pp. 73–77.

CHAPTER 5

_____, DOD 5000.2 – R, Appendix 6, Department of Defense, 5 April 2002.

_____, *DOD Technology Readiness Assessment (TRA) Deskbook*, September 2003, available on-line at http://www.defenselink.mil/ddre/doc/tra_deskbook.pdf

_____, Interim Defense Acquisition Guidebook, Appendix 6, 30 October 2002.

Gold, Rob, "Software Readiness Levels (SWRLs)," presentation, Missile Defense Agency, 2003.

Niemela, Dr. John and Dr. Matthew Fisher, "The Use of Technology Readiness Levels for Software Development," *Army AL&T* magazine, May–June 2004.

Swanson, K., "Technology Readiness Levels Applied to Software," presentation, NASA–Ames Research Center, June 21, 1999.

Turner, Richard, "Status Report on Software Product Maturity Working Group," unpublished, 28 July 2003.

CHAPTER 6

_____, "Bio-Medical Technology Readiness Levels," U.S. Army Medical Research and Materiel Command, Fort Detrick, Maryland, 31 August 2001.

_____, "Infusing Next Generation Learning Systems & Technologies Into University Aerospace Engineering And Science Education," Presentation at the Ad-Hoc Advisory Panel Meeting, NASA Langley Research Center, September 21-22, 2000

_____, "Modeling and Simulation Technologies," Presentation to the Future Combat System Workshop, Department of Energy, National Laboratories, 9–10 February, 2000.

_____, *Technology Readiness Assessment (TRA) Deskbook*, Department of Defense, May 2005.

Broadus, Elizabeth, "Process for Evaluating Logistics Readiness Levels (LRLs)," Presentation to the Multi-Dimensional Assessment of Technology Maturity Conference, 10 May 2006.

Broadus, Elizabeth, "Update on the Process for Evaluating Logistics Readiness Levels (LRLs)," Presentation to the NDIA 9th Annual Systems Engineering Conference, 25 October 2006.

Fiorino, Dr. Thomas D., "Engineering Manufacturing Readiness Levels (EMRLs): Fundamentals, Criteria and Metrics", presentation to OSD/ATL Program Manager's Workshop, 3 June 2003.

Garcia, Suzanne, "TRL Corollaries for Practice-Based Technologies," Presentation to the Multi-Dimensional Assessment of Technology Maturity Conference, 10 May 2006.

Graettinger, Caroline P., PhD, and Suzanne Garcia, "TRL Corollaries for Practice-Based Technologies," Presentation to the Acquisition of Software Intensive Systems Conference, 29 January 2003.

CHAPTER 7

Peisert, Gregory, "Implementing Affordability in Science and Technology," James Gregory Associates, Inc. 2001.

Pyzdek, Thomas, "How Much Don't You Know," *Quality Digest*, February 1999.

Pyzdek, Thomas, *Quality Engineering Handbook*, NY: Marcel Dekker, 2000.

Rickmers, Albert D. and Hollis N. Todd, *Statistics, An Introduction*, New York, McGraw-Hill Publishing Co., 1967.

Shewhart, Walter A., *Economic Control of Quality of Manufactured Product*, New York, D Van Nostrand, 1931.

Taleb, Nassim Nicholas, *The Black Swan: The Impact of the Highly Improbable*, New York, Random House, 2007.

CHAPTER 8

Berra, Yogi, quoted in CBS 60 Minutes, "Famous Yogi-isms," CBS News, 9 July 2003.

Bilbro, James W., "Technology Assessment Requirements for Programs and Projects," presentation at the Multi-Dimensional Assessment of Technology Maturity Workshop, NASA, 10 May 2006.

Bilbro, James W., "Technology Readiness," presentation, NASA, August 2003.

Garvey, P. R., & Cho, C. "An Index to Measure a System's Performance Risk," *Acquisition Review Quarterly*, Vol. 33, No.2, Spring 2003.

Mahafza, Dr. Sherry, Dr. Paul Componation and Dr. Donald Tippett, "A Performance-Based Technology Assessment Methodology to Support DOD Acquisition," *Defense Acquisition Review Journal*, December 2004–March 2005, pp. 268–283.

Mahafza, Dr. Sherry, "Technology Performance Risk Measure," presentation at the Multi-Dimensional Assessment of Technology Maturity Workshop, NASA, 10 May 2006.

Mankins, John C., "Research and Development Degree of Difficulty (R&D3)," NASA White Paper, 1998.

CHAPTER 9

_____, Technology Readiness Assessment (TRA) Deskbook, Department of Defense, May 2005.

Craver, Jeff, "Technology Program Management Model (TPMM) v.2 Brief to AFRL," presentation, US Army Space and Missile Defense Command Technical Center, 2 August 2006.

Craver, Jeff and Mike Ellis, Technology Program Management Model (TPMM) version 2, US Army Space and Missile Defense Command Technical Center, 3 August 2006.

Craver, Jeff and Mike Ellis, "Technology Program Management Model (TPMM) Executive Overview," presentation, US Army Space and Missile Defense Command Technical Center, 29 August 2007.

Taylor, Jack, "DoD Technology Readiness Assessment (TRA) Policy," presentation at the Technology Maturity Conference, OUSD(S&T), 12 September 2007.

© 2006 M. Engling

© 2006

Appendix C

VARIATION AND RISK IN SCIENCE AND TECHNOLOGY

Abstract

Estimating the risk inherent in developing new technologies is important to Air Force Research Laboratory decision makers. Risk can be computed from the variability present in a technology, but the problem is complicated by the absence of data. Sometimes an estimate of variability must be derived from a single data point. This paper examines several techniques that have proved to be useful in attempting to estimate variation as an estimate of standard deviation in a Normal, or Gaussian probability distribution. Results obtained from the various techniques are compared, and we show that one technique has advantages over the others.

Introduction

The Air Force Research Laboratory (AFRL) has found that it is often necessary to make critical investment decisions among technology alterna-

131

tives early during science and technology (S&T) laboratory investigations. Often these decisions must be made in the absence of complete information or even before we have a working prototype demonstrating each alternative. One method for assisting laboratory managers in making these decisions is the value assessment, a technique for quantifying the relative value of competing technical alternatives. The value assessment is based on the AFRL Integrated Product and Process Development (IPPD) process in which both customer satisfaction and risk assessment are used to make effective, and defensible, program planning and execution decisions. The process involves two key metrics, *desirability* and *risk*. *Desirability* is the measurable extent to which a given technology either satisfies or is expected to satisfy a set of requirements. A discussion of desirability is beyond the scope of this paper. *Risk* is measured as the probability that a technology will fail to meet one or more thresholds on those requirements. The threshold is defined as the value that a particular requirement must meet if overall system performance is to be at an acceptable level. Measuring risk means estimating the probability that a technology alternative will fail to meet one or more of our requirements. We say that a technical alternative fails to meet a requirement when the value that the alternative is expected to achieve falls outside of the acceptable range of values (thresholds) established for that requirement.

Figure 1 depicts risk in terms of the portion of a probability distribution (in this case, a Normal, or Gaussian distribution) that lies beyond the threshold. Obviously, if the expected value lies beyond the threshold, risk

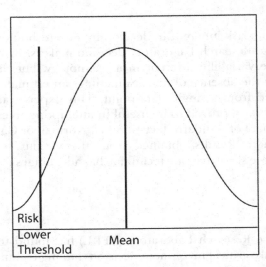

Figure 1.   Gaussian Distribution.

will be high. Even in this case, however, there may be some probability that the actual value achieved in practice might be acceptable. Likewise, when the expected value falls well within the range of acceptability, there still might be significant risk that the actual performance will fall outside the threshold. The reason for these apparently anomalous results comes from the fact that any real process will always have some variation around the expected value. This variation is the source of uncertainty in the reported expected value. This variation is also, therefore, the measure of the risk that, regardless of the expected value, a technical alternative will fail to achieve the threshold.

Risk and variation are so closely related that we can determine risk by estimating variation. If we can find the amount of variation inherent in a system as it relates to a specific requirement, we can easily compute the amount of risk present by calculating the portion of the distribution that falls outside of the threshold. In general, our method for estimating risk consists of finding an estimate for the standard deviation and then using the estimated standard deviation along with the expected value of the variable of interest to define a Normal bell curve. Putting in the value of the threshold then allows us to use a spreadsheet's Cumulative Normal Distribution function to calculate the percent of the distribution that lies outside of the threshold(s). This percentage represents the risk of failing to meet the threshold value for that requirement. Two observations are the immediate result of our method for determining risk:

1. If the expected value (mean) falls exactly on the threshold, risk equals 50% regardless of the estimate of the standard deviation.

2. When the estimate of the standard deviation gets large (approaches infinity), risk approaches 50% regardless of where the expected value falls on the Normal curve.

If we're going to consider variation and risk in the laboratory environment, there must be a compelling reason to do it. After all, AFRL has done fine in the past by simply developing new technologies without quantifying variability and risk. This doesn't mean that we've ignored risk. Many S&T projects included risk mitigation and risk reduction either explicitly or implicitly in the laboratory investigation as well as in the formal program documentation. These risk reduction efforts, however, were done in a qualitative manner, since there were no established statistical methods for quantifying risk in the absence of historical data. Considering risk early is difficult because we don't have the historical database needed to calculate a true standard deviation. In many cases, we may have only a single data point. That single data point might reflect one expert's opinion of the capability of a technical alternative to meet a particular value. The expert's opinion could reflect his/her estimate of what the S&T effort hopes to achieve. The data point might also come from measuring the performance of a single laboratory breadboard or prototype. From that single data point, we're required to derive an estimate of variation if we're to be successful in estimating risk. The problem boils down to finding the answer to this question, "How can I estimate variation from a sample size of one?"

Before we try to answer that question, let's look at an even more basic question, "Why bother?" Why should we try to estimate risk and variation in the S&T environment? Since we're working very early in the development of new technologies, we could just assume that all of our technical alternatives are high-risk ventures. If they're all high risk, measuring risk becomes meaningless since risk is no longer a program discriminator. We have found that there are several reasons for estimating risk in S&T.

Desirability alone doesn't throw out hare-brained or otherwise unworkable technology alternatives. It's possible that a technology alternative could have extremely high desirability, but be so unrealistic as to be worthless. For example, a perpetual motion machine made out of Unobtainium could have high desirability from some viewpoints, such as power requirements. If we ignore the fact that this technology can't be produced, thereby ignoring risk, we could end up embracing a totally unrealistic and unworkable solution. Including an evaluation of risk for each requirement avoids this potential pitfall.

Variation is inherent in all processes. This is as true in research and development as it is in manufacturing. In any endeavor, be it manufacturing or laboratory research, there are sources of variation. Unless we understand the variability inherent in a process, we really don't understand the process very well at all. The earlier we are in the technology life cycle, the more uncertain our knowledge will be, and therefore the greater the potential for variation in the expected outcome. In the S&T environment particularly, our knowledge of the technology and of the production processes necessary for its manufacture is likely to be extremely limited and uncertain. This is shown in Figure 2.

The vertical line represents surety, or certain knowledge. For any given input, we'll know exactly what value the output will have. This is possible only if we totally understand the technology and processes it entails. The narrow area represents a technology where we have a considerable knowledge, and therefore expect a small amount of variability. This is the type of knowledge we would expect of a system undergoing Engineering and Manufacturing Development. But we're in the laboratory phase where the uncertainty and variability is at its most extreme, as demonstrated by the wide area in the figure. Since variation is expected

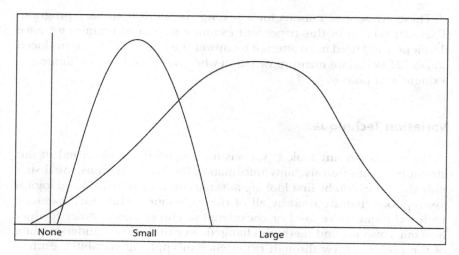

Figure 2.   Degree of Variation.

to be high in the S&T environment, it would be irresponsible to attempt a technology comparison while ignoring this parameter.

We can use variation information as a measure of relative risk. Even though we can assume that all laboratory technology alternatives are risky, it should be obvious that some are riskier than others. Although this fact may be intuitively obvious, it should be possible to quantify this truism. If we can't quantify the degree of risk, we really can't use relative risk as an aid to decision making in S&T investment decisions. If we can develop or estimate information regarding the amount of variability inherent in a given technical alternative, we can use this variation information to compute an indicator of the amount of risk present in that alternative. S&T decision makers will then be able to include an assessment of relative risk in their analysis of the value of competing technologies.

As more and more AFRL decision makers receive training in the AFRL systems engineering initiative and the IPPD process, we find that they expect risk analysis as part of the program documentation. Even our customers (Product centers, Acquisition Offices, and end users) are beginning to demonstrate knowledge of the IPPD process. Some of them expect a value analysis, including a risk assessment, as part of the technology transition plan. In those cases where the customer doesn't specifically require such an analysis and assessment, a formal, quantified evaluation of program risk can be a selling point in planning for technology transition.

These are several reasons for assessing risk during an S&T program. The next section of this paper will examine several techniques we have developed and used in an attempt to answer the other question introduced above, "How can we measure variation when our sample size is limited to a single data point?"

Variation Techniques

We now begin our look at various techniques that we've tried in our attempts to capture variability information for S&T programs. We'll simplify the discussion by first looking at some common statistical and logical assumptions that are used by all of the techniques. Then we'll examine each technique we've used or considered in chronological order to show how our concepts and methods changed over time as our understanding of the problem grew through experience in applying variability estimation to actual data.

Common Assumptions and Calculations

The primary assumption underlying all of the methods and techniques discussed below is that the underlying probability distribution is a Gaussian, or Normal, distribution. We can justify this assumption on three grounds:

1. The Gaussian distribution is well understood and its mathematical behavior is available in the form of readily accessible tables.
2. Many natural processes exhibit Normal behavior.
3. Even when the underlying distribution is non-Gaussian or is unknown, the Central Limit Theorem allows us to use the Normal distribution to describe meaningful statistics drawn from the data under a fairly broad set of conditions.

Making the Gaussian assumption allows us to exploit the well-known characteristics of the bell curve. Several of the methods to be described below make use of the percentage distribution of values around the mean as a function of the number of standard deviations the value is away from the mean (Figure 3). The specific value we often use is that 95.5% of the distribution can be expected to fall within two standard deviations or plus or minus 2σ of the mean. We usually round 95.5% off to 95%, since our data isn't any more accurate than that. If we then make the assumption that the expected range, $x_2 - x_1$, of our data (expected range, since we really have only one data point) represents a 95% confidence interval, this gives the useful relationship $x_2 - x_1 = 4\sigma$. This means that, if we can find

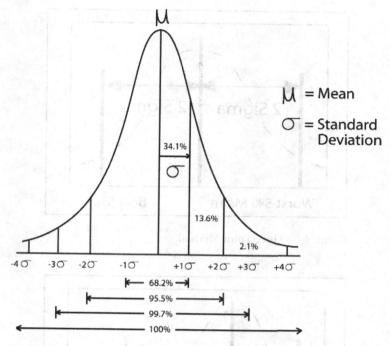

Figure 3. Gaussian Distribution.

an estimate for the range of a variable, we can use this estimate to estimate the standard deviation, σ.

Three-Point (Expected, Best 5%, Worst 5%) Method: The first method we'll look at is the three-point range estimation method. Since the Gaussian distribution is completely described by the mean (μ or Mu) and the standard deviation (σ or Sigma), we can define the entire curve if we know any two of the three values shown as vertical lines in Figure 4. Rounding off the actual values given in Figure 3 by assuming that there are two standard deviations between the Mean (Expected value) and either the Best 5% or Worst 5% lets us calculate any missing values if we know two. For example, given Best and Worst, μ = (Best + Worst)/2 and σ = (Best - Worst)/4. Similarly, if we're given the Expected value and one other value, we already know μ since μ = Mean = Expected value. We can calculate σ by dividing the difference between μ and Best or Worst by 2.

If we can get by with only two values, then why are we asking for three? It's because your Best 5% estimate might not be exactly the same distance from the Expected value as the Worst 5% is. In other words, instead of a perfectly symmetrical Normal distribution, the actual distri-

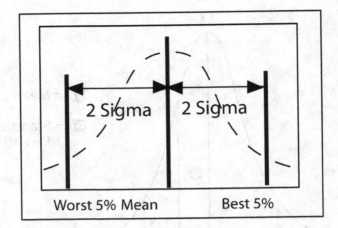

Figure 4.   Three Point Method.

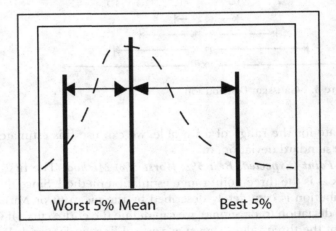

Figure 5.   Skewed Distribution.

bution may be skewed one way or the other. The difference between (Best - μ) and (μ - Worst) will give us a way to estimate the amount of skew. If the difference is 0, we can assume a Gaussian distribution. A value other than 0 will indicate a skewed distribution as shown in Figure 5 above. Bottom line: by collecting three data points, we can gain a better understanding of the total data distribution than is possible with only two.

So what do we mean by "Expected," "Best 5%," and "Worst 5%" values? Think about running 100 completely independent development efforts.

Each effort is run as if the others never existed. They are run concurrently. In some cases, everything goes right. You get lucky. For those programs, you achieve the "Best 5%" of your results. For some programs, Murphy rules and things don't work out. You do not achieve your expected results, and you end up in the "Worst 5%." Somewhere in the rest of the efforts you achieve a median or "Expected" value.

An example of the three point method for some unspecified, "More is better" parameter where we have suppressed the units for simplicity, would be: Given an expected value (E(X)) of 200, "Best 5%" of 240, and "Worst 5%" of 140. We perform the following calculations: (Best − E(X)) = 240 − 200 = 40, and (E(X) - Worst) = 200 − 140 = 60. Although we said above that this difference gives us an idea of skewness, in practice we never used the difference. We simply took the larger value and divided by 2 to get an estimate of σ, S(X) = 60 / 2 = 30.

We used this method to assess risk for several laser technologies. We published a request for information (RFI), and then sent a questionnaire to those companies that responded to the RFI. Although we were able to get useful results, the complexity of the method caused several questionnaire respondents to either ignore the best and worst values, or to assign the same number to all three values. Also, since we only used two of the three values we requested, we sought to find a simpler method to estimate σ.

Bayesian Two-Point Confidence Method: The next method we tried was based on Bayesian statistics as applied to two points in the expected range. The first point was the respondent's expected value. We assigned that point a believability of 95%. We then arbitrarily moved the response variable away from the expected value in the direction of "Better" by approximately 20%, and asked the respondent to estimate their belief that they could achieve this new, better value with their technology. Although the questions were phrased more or less as, "How confident are you that you could achieve this value?" it is important to note that the "belief in ability to achieve" is NOT the same thing as a statistical confidence interval. Here's how it worked in practice: Let the response variable be the random variable X, and an observation of the response variable = x_i, where i = 1, 2. The probability that X = x_i is then estimated from the belief response as follows:

$$P(X = x_i) = \frac{B(X = x_i)}{B(X = x_1) + B(X = x_2)}$$

for i = 1,2. Note that B(X=x_1) is always equal to 0.95 by definition. Also note that performing this calculation for observations x_1 and x_2 will give two probabilities which sum to 100%. Given the probabilities for achieving each response observation, we can then compute a new value for the expected value of X, E(X), as follows:

$$E(X) = x_1 \times P(X = x_1) + x_2 \times P(X = x_2).$$

Finally we can compute an estimate for the standard deviation, S(X), from the probabilities and the calculated expected value:

$$S(X) = \sqrt{x_1^2 \times P(X = x_1) + x_2^2 \times P(X = x_2) - E(X)^2}$$

Given the same situation as we discussed above in the three-point method, the Bayesian two point confidence method calculations give the following results:

Let x_1 = 200; x_2 = 240
Let B(X = x_1) = 0.95; B(X = x_2) = 0.45

$$P(X = x_1) = \frac{B(X = x_1)}{B(X = x_1) + B(X = x_2)} = \frac{0.95}{0.95 + 0.45} = 0.679$$

$$P(X = x_2) = \frac{B(X = x_2)}{B(X = x_1) + B(X = x_2)} = \frac{0.45}{0.95 + 0.45} = 0.321$$

$$E(X) = x_1 \times P(X = x_1) + x_2 \times P(X = x_2)$$

$$E(X) = 200 \times 0.679 + 240 \times 0.321 = 212.9$$

$$S(X) = \sqrt{x_1^2 \times P(X = x_1) + x_2^2 \times P(X = x_2) - E(X)^2}$$

$$S(X) = \sqrt{200^2 \times 0.679 + 240^2 \times 0.321 - 212.9^2} = 18.7$$

The results of these calculations are demonstrated in Figure 6.

This method was used in the analysis of several missile warning technologies. We again published a RFI and sent a questionnaire to the responding companies. Responses to the questionnaire included new technologies as well as off-the-shelf systems. Although we again received useable results in most cases, there was still some confusion evident in the responses. It seems that some people had trouble understanding the

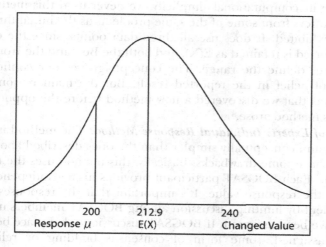

Figure 6.   Bayesian Two-Point Confidence Method.

concept, roughly stated as, "How confident are you in your ability to achieve a value that you didn't provide as your response?" As a result, some simply answered 0%. This response results in a standard deviation estimate of 0, hence no variability and no risk assessment. The idea of arbitrarily computing a new expected value, rather than using the expected value provided in response to the questionnaire, was also hard for some respondents to swallow. This conceptual problem was the major reason driving our search for a simpler method, although the fact that the Bayesian two point confidence method's results are counter-intuitive was also a factor. By this, we mean that as the confidence in the second value increases, so does variation, and vice versa. This result is difficult to explain to individuals with a limited statistical background, hence the confidence that decision makers have in the results is tinged with an attitude of, "Well, if you say so, but..."

Range of Response (Modified Three-Point) Method: This method uses the same three points as the three-point method discussed above. The difference is that, instead of using the maximum value of Best – E(X) and E(X) – Worst, we use the entire range, Best – Worst, and divide by 4. Continuing with the example we've been using, S(X) = (Best – Worst)/4 = (240 – 140)/4 = 25. If we let x_2 = Best and x_1 = Worst, the range is expressed as $x_2 - x_1$, and the estimate for σ then becomes as we demonstrated in Figure 1.

$$S(X) = \frac{x_2 - x_1}{4} = \frac{240 - 140}{4} = 25$$

Despite its computational simplicity, we never used this method. The concept suffers from some of the same problems as the initial three point method exhibited. It does use all three data points, since the expected value reported is retained as E(X), and both the Best and the Worst values are used to define the range. The concept ignores our confidence or strength of belief in the reported result. But the main reason for not using it was that we discovered a new method before the opportunity to apply this method arose.

Board of Experts Individual Response Method: The method is computationally and conceptually simpler than the ones described above, but it does still have some drawbacks. Basically, this method uses the BOGSAT approach.[1] Each BOGSAT participant provides his/her independent estimate for the response value. It's important that the responses be kept independent by limiting discussion among BOGSAT members until after values have been assigned. If BOGSAT discussion takes place before values are assigned, some form of consensus building or reliance on BOGSAT authority figures may dilute the variation in the individual responses. The individual responses are then used to calculate a mean (μ) and a standard deviation (σ) using standard statistical formulae to perform the computations.

The main advantage of this method is its conceptual simplicity. It's easier to explain how μ and σ can be calculated from a collection of expert opinions than it is to explain the systems outlined above. Also, since all values obtained from each expert are retained, the BOGSAT members can have substantial disagreement on values assigned, and since there is no need for consensus, the disagreements don't have to be resolved.

This last advantage is also a potential disadvantage. Since there is no consensus, BOGSAT members with uninformed opinions may unduly influence the calculated values for both μ and σ.

Another disadvantage of the method is that if all BOGSAT members agree on a single response value we still have the problem of trying to determine variation for that requirement. Because of this, the BOGSAT approach requires a large enough pool of experts to derive meaningful statistics from their individual opinions. It may not always be possible to find a group of people who have sufficient depth of knowledge to develop individual estimates of the capabilities of a particular technical alternative. Even if such a group of experts can be identified, the amount of time required for each to work out his/her own independent responses in situations where there might be dozens of individual requirements can be prohibitive.

The method was simulated for this paper by separately asking a group of ten individuals to give a number between 100 and 350. The responses were as follows: 267, 112, 272, 217, 140, 275, 176, 220, 198, and 150. The statistics of this sample data are a mean, E(X) of 202.7 and a standard deviation, S(X), of 58.06.

We never used this method as stated; however, we did use a variation of the BOGSAT method to establish a consensus on the requirements, thresholds, and objectives for several technology analyses.

Believability/Height of Gaussian Curve Method: It is possible to compute a value for the standard deviation of an assumed Normal (Gaussian) distribution by estimating the height of the bell-shaped curve. Given that the area under the curve represents a cumulative probability, that total area for the normalized Gaussian distribution ($\mu = 0$, $\sigma = 1$) must always be equal to 1.00. As the height of the curve increases, the curve must become more peaked to maintain the constant area. Conversely, as the height of the curve decreases, the curve must flatten out. This is shown in Figure 7. The numerical measure of this flatness is the standard deviation, σ, which is described as a measure of variation around the expected value, or mean (μ). A large value of σ results in a flatter curve with more variation than a curve that has a smaller σ. As a result, the curve with a small σ will be more peaked than the other one. In the limit, as σ goes to 0, the curve becomes a δ function or spike ($\uparrow$) of zero width with height equal to infinity and area of 1. As the height goes to 0, the curve becomes a horizontal straight line at zero height from $-\infty \rightarrow \infty$, again with area under the curve equal to 1.

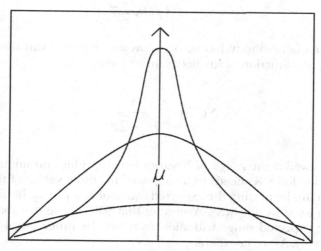

Figure 7.   Flatness vs. Height of Gaussian Curve.

We can use the equation for the Gaussian distribution to find σ as a function of the height of the curve at the point where the random variable X is equal to the mean, or x = μ. The Gaussian equation is

$$F(x) = \frac{1}{\sigma\sqrt{2\pi}} e^{-\frac{(x-\mu)^2}{2\sigma^2}}$$

where F(x) is the height of the curve at x. If we try to solve for σ, we have a real mess; however, letting x = μ simplifies the expression enormously. At x = μ, the equation becomes

$$F(\mu) = \frac{1}{\sigma\sqrt{2\pi}} e^{0}$$

or simply

$$F(\mu) = \frac{1}{\sigma\sqrt{2\pi}} .$$

This is easily solved for σ,

$$\sigma = \frac{1}{F(\mu)\sqrt{2\pi}} .$$

If we let the believability factor, B_μ, represent F(μ), we can find an estimate of σ, as a function of the believability factor,

$$S(X) = \frac{1}{B_\mu \sqrt{2\pi}} .$$

This works well if the range of X varies between plus and minus infinity, but the value for S(X) needs to be adjusted for finite values of the range. We can do this by putting the expected range of X, x_1 to x_2, in the numerator when we solve for S(X). Assuming that 95% of the distribution falls within the expected range will allow us to set the range equal to 4 S(X) (See Figure 4). As we saw there,

$$S(X) = \frac{x_2 - x_1}{4}.$$

and from immediately above,

$$S(X) = \frac{1}{B_\mu \sqrt{2\pi}}.$$

Multiplying these two expressions for S(X) then gives us

$$S(X)^2 = \frac{x_2 - x_1}{4} \times \frac{1}{B_\mu \sqrt{2\pi}}.$$

We take the square root of the right hand term of the equation to get S(X). Our equation expressing the estimate for the standard deviation, σ, as a function of our belief that the expected value, μ, is correct then becomes

$$S(X) = \sqrt{\frac{x_2 - x_1}{4B_\mu \sqrt{2\pi}}}$$

Using this method on our example gives the following:

Range = $x_2 - x_1$ = 240 – 140 = 100
Believability factor B_μ = 0.95

$$S(X) = \sqrt{\frac{x_2 - x_1}{4B_\mu \sqrt{2\pi}}} = \sqrt{\frac{240 - 140}{4 \times 0.95 \times \sqrt{2\pi}}} = 3.24$$

This method allows us to develop an estimate for the standard deviation in a much simpler manner than the methods previously used. Of all the methods we looked at, this one is by far the easiest to explain. With an estimate of the mean and our belief in the accuracy of the estimate, we can compute an estimate for the standard deviation. This estimate can then be used to quantify risk.

Having found an estimate for the standard deviation, we can then compute risk by finding the portion of the curve that falls beyond the established threshold. Typically, spreadsheet programs have built-in functions that perform this calculation for both the one-tailed (More is better or less is better) case (Figure 8) or two-tailed (Target value) case.

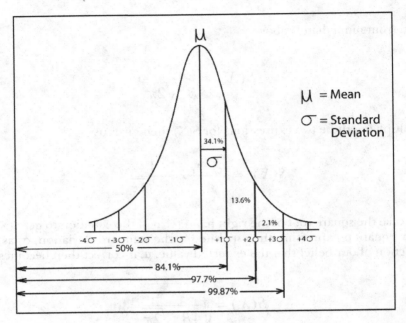

Figure 8.   One-Tailed Gaussian Distribution.

One problem with the believability vs. height method is its behavior near the extreme upper values. When believability is high, as it is in this case, we get a fairly low estimate of σ, in the example, 3.24. The problem shows itself when we put believability at 100%. When we're absolutely sure of our answer, σ ought to be equal to 0, not just low. The problem is caused by the fact that when we put believability equal to 1, the height of the curve is assumed to be 1 in our formula. Actually, for a believability of 1, the height of the curve should be infinite. We can modify the formula by including a function that goes from 0 to infinity as believability goes from 0 to 1. The tangent function behaves in the desired fashion, but to use the tangent, we need to multiply the believability factor by π/2. Our equation now is

$$S(X) = \sqrt{\frac{x_2 - x_1}{4\tan\left(\frac{\pi}{2}B_\mu\right)\sqrt{2\pi}}}$$

Let's see how this equation behaves. If $B_\mu = 1$, $\tan \pi/2 = \infty$ and $S(X) = 0$. As we decrease the believability factor, σ increases, until at a believability value near the lower extreme of only 1% ($B_\mu = 0.01$), σ becomes 25.2 and it increases rapidly thereafter until, at $B_\mu = 0$, σ becomes infinite.

We used this method to evaluate technology concepts for a laboratory program that was very early in development. We were looking mostly at paper studies, with little or no laboratory experimental data available. There was virtually no hardware to test, not even at the prototype or breadboard level of maturity. That's why we used this method.

Modified Believability/Height of Gaussian Curve Method: The Believability/Height of Gaussian Curve Method described above has a problem at the low end. As believability goes to 0, the tangent function also goes to 0, $\mathrm{Tan}(0) = 0$. Although this is the value we're looking for in the lower limit as believability goes to 0, it would be more esthetically pleasing and more symmetric if the chosen function approached 0 asymptotically, just as it approaches infinity asymptotically on the high end. We can arrange this by using the full range of the tangent function by letting the believability vary from $-\pi/2$ to $\pi/2$ instead of from 0 to 1.

This solution, however, introduces an additional complication since the value of the tangent function goes from $-$ to $+$ rather than from 0 to $+$ on the $-\pi/2$ to $\pi/2$ interval. We can fix this defect by using the exponential function. As we raise e to increasingly large negative powers, the function approaches 0 asymptotically, just as we wish. At the upper limit, e raised to a large positive power still approaches infinity, so we have the behavior we seek. We achieve this result by introducing another level of complexity, since our equation for the estimated standard deviation now looks like this:

$$S(X) = \sqrt{\frac{x_2 - x_1}{4e^{\tan\left(\pi B_\mu - \frac{\pi}{2}\right)}\sqrt{2\pi}}}$$

SUMMARY AND COMPARISONS

We've looked at several techniques for estimating variation from limited data. The methods discussed here are by no means exhaustive, but they do demonstrate that the problem of variation estimation is, in fact, solvable.

The preferred estimation technique, Modified Believability/Height of Gaussian Curve method, gives virtually no risk for a high believability factor and high risk for low believability. This result matches the gut feel of how the underlying statistics ought to behave.

We have shown, although without mathematical rigor, that it is possible to derive an estimate for statistical variation when there is extremely limited data available. There are a number of useful techniques that can be used to estimate variation from a sample size of one.

Table of Results: Table 1 presents a capsule summary of the results for the sample inputs for the problem as given above. A blank cell means that value isn't used for a particular technique.

The bottom row in the table shows how the variability computations for each technique result in different estimates for the amount of risk present if the threshold value is pegged at 180. Note the wide variation in the results. This method dependent variation demonstrates that each of these techniques gives only an estimate of the true variability. This limitation may not be critical for a simple rank-ordering problem, since all potential solutions evaluated will suffer from the same inaccuracies inherent in a given technique. Even if all of the methods converge to a single value, however, the results of this estimation exercise should not be used as accurate representations of statistical probabilistic risk assessments.

Table 1.   Results

	Three-Point	Bayesian	Range	BOGSAT	Believability	Modified Believability
y_1	200	200	200		200	200
$B(Y = y_1)$		0.95			0.95 0.05	0.95 0.05
B_μ					0.001	0.001
y_2		240				
$B(Y = y_2)$		0.45				
x_1	140		140	112	140	140
x_2	240		240	275	240	240
Range	100		100		100	100
Best $-\mu$	40					
μ - Worst	60					
$E(Y)$	200	212.9	200	202.7	200	200
$S(Y)$	30	18.7	25	58.06	0.89 11.26 79.68	0.13 74.21 4.16E69
Threshold	180	180	180	180	180	180
Risk (ζ)	25.25%	3.93%	21.19%	34.79%	0.00% 3.78% 40.09%	0.00% 39.38% 50.0%

NOTE

1. BOGSAT means, "Bunch Of Guys Sitting Around a Table." For political correctness, the acronym must be understood to mean guys in the generic, inclusive, non-gender specific sense. Alternatively, "Guys" could be replaced by "Geeks," but the term, "Geeks" might carry some unintended pejorative emotional content. The G in the acronym could also stand for "Guys and Gals," but that term may not be politically correct in some settings. The best solution might be to change the acronym to BOPSAT for "Bunch Of People Sitting Around a Table," but BOGSAT has become enshrined as a more or less standard acronym in some circles, and the term may not be amenable to change.

© 2006 M. Engling

D

© 2006

SYSTEMATIC ASSESSMENT OF THE PROGRAM/PROJECT IMPACTS OF TECHNOLOGICAL ADVANCEMENT AND INSERTION

James W. Bilbro
Assistant Director for Technology/Chief Technologist
Office of the Director
George C. Marshall Space Flight Center
Huntsville, AL 35801
December, 2006

Acknowledgments: I would like to acknowledge and express my appreciation for the valuable contributions to the preparation of this paper by the following individuals: Dale Thomas, Jack Stocky, Dave Harris, Jay Dryer, Bill Nolte, James Cannon, Uwe Hueter, Mike May, Joel Best, Steve

151

Newton, Richard Stutts, Wendel Coberg, Pravin Aggarwal, Endwell Daso, Bill Burt, and Steve Pearson.

EXECUTIVE SUMMARY

Technology development plays a far greater role in the life cycle of a program/project than has been traditionally considered and it is the role of the systems engineer to develop an understanding of the extent of program/project impacts - maximizing benefits and minimizing adverse effects. Traditionally, from a program/project perspective, technology development has been associated with the development and incorporation of any "new" technology necessary to meet requirements. However, a frequently overlooked area is that associated with the modification of "heritage" systems incorporated into different architectures and operating in different environments from the ones for which they were designed. If the required modifications and/or operating environments fall outside of the realm of experience then this too should be considered technology development.

In order to understand whether or not technology development is required—and to subsequently quantify the associated cost, schedule and risk, it is necessary to systematically assess the maturity of each system, sub-system or component in terms of the architecture and operational environment. It is then necessary to assess what is required in the way of development to advance the maturity to a point where it can be successfully incorporated within cost, schedule and performance constraints. Because technology development has the potential for such significant impacts on a program/project, Technology Assessment needs to play a role throughout the design and development process from concept development through Preliminary Design Review (PDR). Lessons learned from a technology development point-of-view should then be captured in the final phase of the program.

Stakeholder Expectation: Government Accountability Office (GAO) studies have consistently identified the "mismatch" between stakeholder expectation and developer resources (specifically the resources required to develop the technology necessary to meet program/project requirements) as a major driver in schedule slip and cost overrun.

Requirements Definition: If requirements are defined without fully understanding the resources required to accomplish needed technology developments then the program/project is at risk. Technology assessment must be done iteratively until requirements and available resources are aligned within an acceptable risk posture.

Design Solution: As in the case of requirements development, the design solution must iterate with the technology assessment process to ensure that performance requirements can be met with a design that can be implemented within the cost, schedule and risk constraints.

Risk Management: In many respects, technology assessment can be considered a subset of risk management and as such should be a primary component of the risk assessment. A stand alone report of technology readiness assessment must be provided as a deliverable at PDR per NPR 7120.5d.

Technical Assessment: Technology assessment is also a subset of technical assessment and implementing the assessment process provides a substantial contribution to overall technical assessment.

Trade Studies: Technology assessment is a vital part of determining the overall outcome of Trade Studies, particularly with decisions regarding the use of heritage equipment.

Verification/Validation: The verification/validation process needs to incorporate the requirements for technology maturity assessment in that in the end maturity is demonstrated only through test and/or operation in the appropriate environment.

Lessons Learned: Part of the reason for the lack of understanding of the impact of technology on programs/projects is that we have not systematically undertaken the processes to understand impacts.

INTRODUCTION, PURPOSE, AND SCOPE

The Agency's programs and projects, by their very nature, frequently require the development and infusion of new technological advances in order to meet performance requirements arising out of mission goals and objectives. Frequently, problems associated with technological advancement and subsequent infusion have resulted in schedule slips, cost overruns and occasionally even to cancellations or failures. It is the role of the Systems Engineer to develop an understanding of where those technological advances are required and to determine their impact on cost, schedule and risk. It should be noted that this issue is not confined to "new" technology. Often major technological advancement is required for a "heritage" system that is being incorporated into a different architecture and operated in a different environment from that for which it was originally designed. In this case, it is frequently not recognized that the adaptation requires technological advancement and as a result, key systems engineering steps in the development process are given short shrift—usually to the detriment of the program/project.

In both contexts of technological advancement (new and adapted heritage), infusion is a very complex process that has been dealt with over the years in an ad hoc manner differing greatly from project to project with varying degrees of success. In post mortem, the root cause of such events has often been attributed to "inadequate definition of requirements." If such were indeed the "root cause," then correcting the situation would simply be a matter of requiring better requirements definition, but since history seems frequently to repeat itself, this must not be the case—at least not in total. In fact there are many contributors to schedule slip, cost overrun, project cancellation and failure—among them lack of adequate requirements definition. The case can be made that most of these contributors are related to the degree of uncertainty at the outset of the project and that a dominant factor in the degree of uncertainty is the lack of understanding of the maturity of the technology required to bring the project to fruition and a concomitant lack of understanding of the cost and schedule reserves required to advance the technology from its present state to a point where it can be qualified and successfully infused with a high degree of confidence, in other words, where requirements and available resources are in line. Although this uncertainty cannot be eliminated, it can be substantially reduced through the early application of good systems engineering practices focused on understanding the technological requirements, the maturity of the required technology and the technological advancement required to meet program/project goals, objectives and requirements. The only way to ensure the necessary level of understanding is for the systems engineer to conduct a systematic assessment of all systems, subsystems and components at various stages in the design/ development process. It is extremely important to begin the assessment at the earliest possible point since results play a major role in the determination of requirements; the outcome of trade studies; the available design solutions; the assessment of risk; and the determination of cost and schedule,

There are a number of processes that can be used to develop the appropriate level of understanding required for successful technology insertion. None of them provide the complete answer, but it is the intent of this section is to describe a systematic process that can be used as an example of how to apply standard systems engineering practices to perform a comprehensive Technology Assessment (TA) that will go a long way toward reducing the level of uncertainty in program/project success. It should be noted that the examples provided in this section are just examples. The process can and should be tailored to meet the needs of the particular program/project to which it is being applied.

The TA is comprised of two parts, a Technology Maturity Assessment (TMA) and an Advancement Degree of Difficulty Assessment (AD^2). The process begins with the TMA which is used to determine technological maturity via NASA's Technology Readiness Level (TRL) scale. It then proceeds to develop an understanding of what is required to advance the level of maturity through a process called the Advancement Degree of Difficulty (AD^2).

It is necessary to conduct a TA at various stages throughout a program/project in order to provide the Key Decision Point (KDP) products required for transition between phases identified below:

KDP A—Transition from Pre-Phase A to Phase A:

- Requires an assessment of potential technology needs versus current and planned technology readiness levels, as well as potential opportunities to use commercial, academic, and other government agency sources of technology. Included as part of the draft integrated baseline.

KDP B—Transition from Phase A to Phase B:

- Requires a Technology Development plan identifying technologies to be developed, heritage systems to be modified, alternate paths to be pursued, fall back positions and corresponding performance descopes, milestones, metrics and key decision points. Incorporated in the preliminary Project Plan.

KDP C—Transition from Phase B to Phase C/D:

- Technology Readiness Assessment Report (TRAR) demonstrating that all systems, subsystems and components have achieved a level of technological maturity with demonstrated evidence of qualification in a relevant environment.

The initial TMA serves to provide the baseline maturity of the system at program/project outset, and to monitor progress throughout development. The final TMA is performed just prior to the PDR and it forms the basis for the Technology Readiness Assessment Report (TRAR) which documents the maturity of the systems, subsystems and components demonstrated through test and analysis. The initial AD^2 assessment provides the material necessary to develop preliminary cost and schedule plans and preliminary risk assessments. In subsequent assessment, the information is then used to build the technology development plan, in the process identifying alternative paths; fall-back positions and performance descope options. The information is also vital to the preparation of milestones and metrics for subsequent Earned Value Management.

The assessment is performed against the hierarchical breakdown of the hardware and software products of the program/project Work Breakdown Structure (WBS) in order to achieve a systematic, overall understanding at the system, subsystem and component level (Figure 1).

INPUTS/ENTRY CRITERIA

It is extremely important that a Technology Assessment process be defined at the beginning of the program/project and that it be performed at the earliest possible stage (concept development) throughout the program/project through PDR. Inputs to the process will vary in level of detail according to the phase of the program/project, and even though there is a lack of detail in pre-phase A, the TA will drive out the major critical technological advancements required. Therefore, at the beginning of pre-phase A, the following should be provided:

- Refinement of Technology Readiness Level Definitions
- Definition of Advancement Degree of Difficulty Process
- Definition of terms to be used in the assessment process
- Establishment of meaningful evaluation criteria and metrics that will allow for clear identification of gaps and shortfalls in performance.
- Establishment of the TA team
- Establishment of an independent TA review team.

HOW TO DO A TECHNOLOGY ASSESSMENT

The technology assessment process makes use of basic systems engineering principles and processes. As mentioned previously, it is structured to occur within the framework of the WBS in order to facilitate incorporation of the results. Using the WBS as a framework has a two-fold benefit—it breaks the "problem" down into systems, subsystems and components that can be more accurately assessed, and it provides the results of the assessment in a format that can readily be used in the generation of program costs and schedules. It can also be highly beneficial in providing milestones and metrics for progress tracking using Earned Value Management. An example is shown in Figure 1.

As discussed above, Technology Assessment is a two step process comprised of step one, the determination of the current technological maturity in terms of Technology Readiness Levels (TRL's) and step two; the

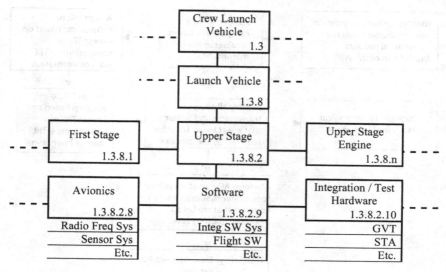

Figure 1. WBS Example.

determination of the difficulty associated with a moving a technology from one TRL to the next through the use of the Advancement Degree of Difficulty (AD^2). The overall process is iterative, starting at the conceptual level during program formulation, establishing the initial identification of critical technologies and the preliminary cost, schedule and risk mitigation plans. Continuing on into Phase A, it is use to establish the baseline maturity, the Technology Development plan and associated costs and schedule. The final TA consists only of the TMA and is used to develop the TRAR which validates that all elements are at the requisite maturity level (see Figure 2).

Even at the conceptual level, it is important to use the formalism of a WBS to avoid having important technologies slip through the crack. Because of the preliminary nature of the concept, the systems, subsystems and components will be defined at a level that will not permit detailed assessments to be made. The process of performing the assessment, however, is the same as that used for subsequent, more detailed steps that occur later in the program/project where systems are defined in greater detail.

Once the concept has been formulated and the initial identification of critical technologies made, it is necessary to perform detailed architecture studies with the Technology Assessment Process intimately interwoven (Figure 3). The purpose of the architecture studies is to refine end-item system design to meet the overall scientific requirements of

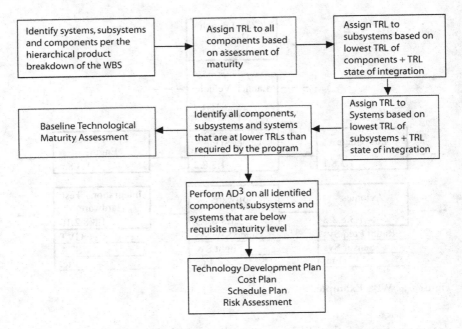

Figure 2.   Technology Assessment Process.

the mission. It is imperative that there be a continuous relationship between architectural studies and maturing technology advances. The architectural studies must incorporate the results of the technology maturation, planning for alternate paths and identifying new areas required for development as the architecture is refined. Similarly, it is incumbent upon the technology maturation process to identify requirements that are not feasible and development routes that are not fruitful and to transmit that information to the architecture studies in a timely manner. The architecture studies in turn must provide feedback to the technology development process relative to changes in requirements. Particular attention must be given to "heritage" systems in that they are often used in architectures and environments different from those in which they were designed to operate.

ESTABLISHING TECHNOLOGY READINESS LEVELS

TRL is, at its most basic, a description of the "performance history" of a given system, subsystem or component relative to a set of levels first described at NASA Headquarters in the 1980's. The TRL essentially describes the level of maturity of a given technology and provides a "base-

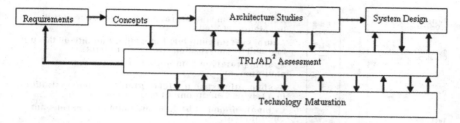

Figure 3.   Architectural Studies and Technology Development.

line" from which maturity is gauged and advancement defined. Even though the concept of TRL has been around for almost 20 years, it is not well understood and frequently misinterpreted. As a result we often undertake programs without fully understanding either the maturity of key technologies or what is needed to develop them to the required level. *It is impossible to understand the magnitude and scope of a development program without having a clear understanding of the baseline technological maturity of all elements of the system.* Establishing the TRL is a vital first step on the way to a successful program. A frequent misconception is that in practice it is too difficult to determine TRLs and that when you do it is not meaningful. On the contrary, identifying TRLs can be a straightforward systems engineering process of determining what was demonstrated and under what conditions was it demonstrated.

At first blush, the TRL descriptions in Figure 4 appear to be straight forward. It is in the process of trying to assign levels that problems arise. A primary cause of difficulty is in terminology—everyone knows what a breadboard is, but not everyone has the same definition. Also, what is a "relevant environment?" What is relevant to one application may or may not be relevant to another. Many of these terms originated in various branches of engineering and had, at the time, very specific meanings to that particular field. They have since become commonly used throughout the engineering field and often take differences in meaning from discipline to discipline, some subtle, some not so subtle. Breadboard for example comes from electrical engineering where the original use referred to checking out the functional design of an electrical circuit by populating a "breadboard" with components to verify that the design operated as anticipated. Other terms come from mechanical engineering, referring primarily to units which are subjected to different levels of stress under testing, i.e. qualification, proto-flight and flight units. The first step in developing a uniform TRL assessment is then to define the terms used. An example set of definitions appears in Attachment A and a further refinement of the TRL scale to include software is shown in Attachment B.

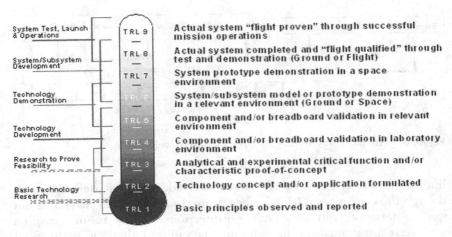

Figure 4.   Technology Readiness Levels.

It is extremely important to develop and use a consistent set of definitions over the course of the program/project.

Having established a common set of terminology, it is necessary to proceed to the next step—quantifying "judgment calls" on the basis of past experience. Even with clear definitions there will be the need for "judgment calls" when it comes time to assess just how similar a given element is relative to what is needed (i.e., is it "close enough" to a prototype to be considered a prototype, or is it more like an engineering breadboard?) Describing what has been done in terms of form, fit and function provides a means of quantifying an element based on its design intent and subsequent performance. Figure 5 provides a graphical representation of a 3-dimensional continuum where various models, breadboards, engineering units, prototypes, etc. are plotted. The X-axis of the graph represents "function," the Y-axis represents "form" and the Z-axis represents "fit." A breadboard, according to our definition, demonstrates function only without regard to form or fit and would consequently fall on the X-axis. If the breadboard was of the full "system," then it would be at the 100% mark; if instead it was a breadboard of a subsystem or component, it would fall somewhere to the left of the 100% mark.

Another example would be that of a wind tunnel model. These models are typically less than 1% scale, but demonstrate aerodynamic properties associated with "form." A mass model could demonstrate both form and fit, but would have no function. The prototype would be as close to the final article as possible and would consequently demonstrate form, fit and function. By plotting the device under question, it will be easier to classify its state of development.

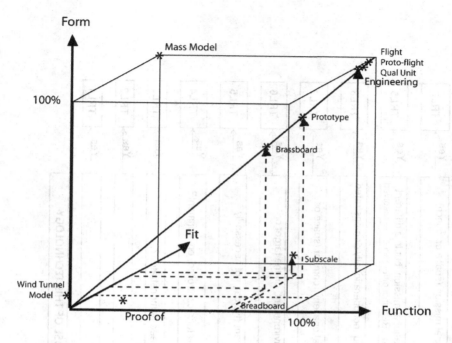

Figure 5. Form, Fit, and Function.

A third critical element of any assessment relates to the question of who is in the best position to make "judgment calls" relative to the status of the technology in question. For this step, it is extremely important to have a well-balanced, experienced assessment team. Team members do not necessarily have to be "discipline experts," but they do have to have a good understanding at the system, or subsystem level of what has been done under what type of conditions and how that relates to what is under evaluation. Establishing a team having the appropriate level of experience is the most critical aspect of technology assessment.

Having established a set of definitions, defined a process for quantifying "judgment calls," and assembled an expert assessment team, the process primarily consists of asking the right questions. The flow chart depicted in Figure 6 demonstrates the questions to ask in order to determine TRL at any level in the assessment.

Note the 2nd box particularly refers to heritage systems. If either the architure or the environment in which the system is operating has changed from that in which it was originally operating, then the TRL for that system

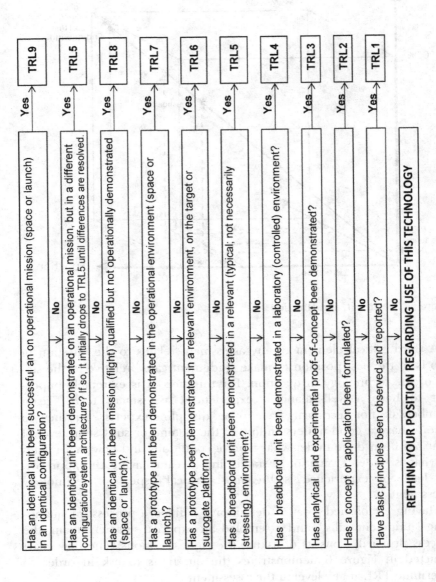

Figure 6. The TMA Assessment Thought Process.

drops to TRL 5—at least initially. If in subsequent analysis the new environment is sufficiently close to the old environment, or the new architecture sufficiently close to the old architecture, the resulting evaluation could be TRL 6 or 7. The most important thing to realize is that it is no longer at a TRL 9. Applying this process at the system level and then proceeding to lower levels of subsystem and component identifies those elements that require development and sets the stage for the subsequent phase, determining the AD^2. A method for formalizing this process is shown in Figure 7. In Figure 7, the process has been set up as a table, on the left hand side of the table the rows identify the systems, subsystems, and components that are under assessment. The columns identify the categories that will be used to determine the TRL, i.e., what units have been built, to what scale, and in what environment have they been tested? Answers to these questions determine the TRL of an item under consideration. The TRL of the system is determined by the lowest TRL present in the system, i.e., a system is at TRL 2 if any single element in the system is at TRL 2. The problem of multiple elements being at low TRLs is dealt with in the AD^2 process. It should be noted that the issue of integration affects the TRL of every system, subsystem and component. All of the elements can be at a higher TRL, but if they have never been integrated as a unit, the TRL will be lower for the unit. How much lower depends on the complexity of the integration.

TRL Assessment										
		Demo Units			Environment			Description			TRL
System / Subsystem	Concept	Breadboard	Brassboard	Prototype, etc.	Laboratory	Relevant	Space, etc.	Form	Fit	Function, etc.	Overall TRL
1.0 System											2
1.1 Subsystem x											2
1.1.1 Mechanical Comp		X			X						2
1.1.2 Mechanical Sys			X			X				X	4
1.1.3 Electrical Comp			X	X		X		X	X	X	6
1.1.n Etc.	X										
1.2 Subsystem y											5
1.2.1 Electrical Sys	X				X	X		X		X	6
1.2.n Etc.											5
1.n Etc.											

Figure 7.   TRL Assessment Matrix.

TRL CALCULATOR

An automated TRL calculator which incorporates all of these issues (and more) has been developed and is available for use upon request. The process flow for the calculator is shown in Figure 8. The calculator can be set up to address TRL's or MRL's (Manufacturing Readiness Levels) as is shown in Figure 9. The assessment summary is shown in Figure 10, and an example of an assessment is shown in Figure 11.

ASSESSING THE AD^2 IN MOVING TECHNOLOGY TO A HIGHER TRL

Once a TRL has been established for the various elements of the system, subsystem or component under development, it becomes necessary to assess what will be required to advance the technology to the level required by the program. The importance of this step cannot be overemphasized because everything that transpires from this point forward will depend upon the accuracy of the assessment: the ability to prepare realistic schedules, make realistic cost projections, meet milestones and

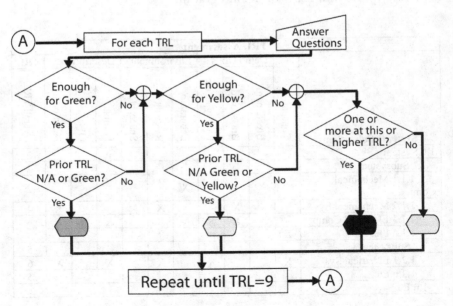

Figure 8.   The TRL Calculator Process Flow.

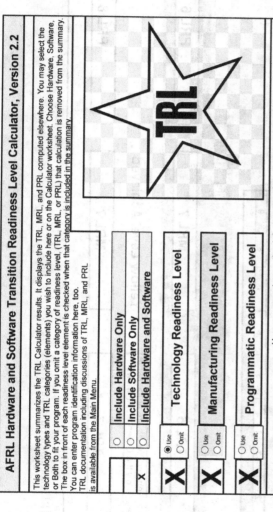

Figure 9.   The TRL Calculator Option Selection.

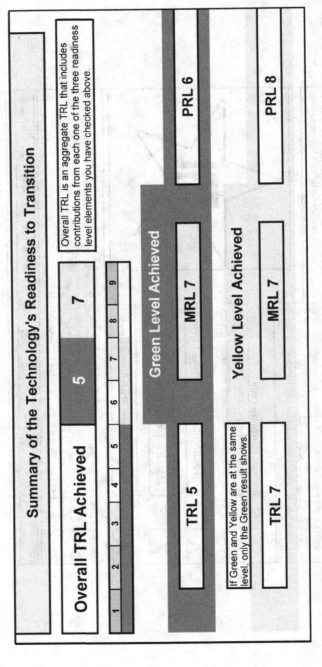

Figure 10. TRL Calculator Summary Page.

WBS	Element Name	Assessed Area	TRL Score			MRL Score		
			Grn	Yel	Red	Grn	Yel	Red
136905.08.05.03	Structure & Thermal	Primary structure	5	6	7	5		
		Secondary structure	5	6	7	5	6	
		Flight Separation System	5	6, 7	8	5	7	
		TPS	5	6	8	6	7	
136905.08.05.04	Main Propulsion System	Ducts & Lines	5	6	7	5	6	
		Valves & Actuators	5	6, 7	7	5	6	
136905.08.05.05	US Reaction Control System		5	6, 7	8	5	7	
136905.08.05.06	1st Stage Reaction Control System		5	6	8	5	7	
136905.08.05.07	Thrust Vector Control	Actuators	5	6	7	5	6	
		Hydraulic Power Systems	4	5, 6	7	5	6	
136905.08.05.08	Avionics	C&DH	4	5, 6	7	5	6	
		GNC Hardware	5	6	7	6		
		Electrical Power	6	7		6	7	
		Sensors	6	7		6	7	

Figure 11. TRL Calculator Results.

ultimately produce the desired results all depend upon having as accurate AD^2 assessment as possible. *This assessment is one of the most challenging aspects of technology development—not all technologies are the same.* It requires the art of "prediction," and the accuracy of the prediction relies on:

- Expert personnel
- Detailed examination of required activity.
- Review by independent advisory panel

Establishing an accurate AD^2 is difficult, but if approached in a systematic manner such as that used for establishing the TRL, then it becomes tractable. Once again, the key is to break it down to a fine enough granularity that problems can be quantified and solution probabilities accurately estimated. The AD^2 of the overall system under development will be a function of the AD^2 of the individual subsystems and components. It is not a straightforward combination of AD^2 s. Neither is it determined solely by the most difficult element (although basing an entire system around a key component that is made of "unobtainium" would definitely place the entire system in the "breakthrough physics" category). A more likely case is where some of the elements have been demonstrated at the appropriate level, some are relatively close to being demonstrated and some have some significant hurdles to overcome. The more elements you have with low AD^2 values, the greater the difficulty in advancing the system as a whole to the requisite level. This is intuitively obvious, however, knowing that a large number of elements have low AD^2 values means that the problem has been examined at sufficient granularity. This is a difficult and time-consuming process that requires a broad mix of highly experienced individuals, and it is a process that is often inadequately performed. Simply put, a cursory examination of a problem may determine that there is only one element with a low AD^2 value when in fact there are two or more such elements. If the program is structured to attack the identified area, it may in fact resolve the problem, only to find that the program is brought to a halt by one or more of the remaining problems.

Early recognition of the need for "quantifying" the difficulty in maturing technology resulted in establishing five categories of difficulty. The AD^2 expands this concept into a systems engineering approach to systematically evaluate the effort required to advance the technological maturity of the elements identified in the TRL Assessment as being insufficiently mature. It combines a number of processes, Integration Readiness Levels, System Readiness Levels, Manufacturing Readiness Levels, Design Readiness Levels, Materials Readiness Levels, etc. into one overarching process. The AD^2 increases the number of levels of difficulty from five to nine and reorganizes them, in recognition of a need for greater distinction in difficulty and the desire to have it more in line with the structure of the TRL scale (Table 1)

In determining the AD^2, the focus is on the tools, processes and capabilities associated with the categories that address the development process required to produce a given element; i.e., design, analysis, manufacturing, operation, test and evaluation. It also includes categories of operations that must be considered in the development process and the development stages that must be undertaken to reduce risk. The result of the AD^2 process is a matrix that provides a quantifiable assessment of the degree of difficulty in advancing the technology. Figure 12 is an example showing the process in assessing the status of turbomachinery, it involves the systems, subsystems and components identified in the maturity assessment phase as ones that were insufficiently mature. The left hand side of the table is identical in form to that of the TRL assessment table but may in fact go to even finer granularity. The identification of the columns, however, is quite different. In this case, we are not looking for the existence of historical information relative to the current status of an element, rather we are looking for an assessment of the degree of difficulty in maturing the element. This requires addressing appropriate questions regarding the development process:

Table 1. Advancement Degree of Difficulty Level Descriptions

Degree of Difficulty	Description
9	0% Development Risk—Exists with no or only minor modifications being required. A single development approach is adequate.
8	10% Development Risk—Exists but requires major modifications. A single development approach is adequate.
7	20% Development Risk—Requires new development well within the experience base. A single development approach is adequate.
6	30% Development Risk—Requires new development but similarity to existing experience is sufficient to warrant comparison across the board. A single development approach can be taken with a high degree of confidence for success.
5	40% Development Risk—Requires new development but similarity to existing experience is sufficient to warrant comparison in all critical areas. Dual development approaches should be pursued to provide a high degree of confidence for success.
4	50% Development Risk - Requires new development but similarity to existing experience is sufficient to warrant comparison in only a subset of critical areas. Dual development approaches should be pursued to provide a moderate degree of confidence for success. (Desired performance can be achieved in subsequent block upgrades with high degree of confidence.)
3	60% Development Risk—Requires new development but similarity to existing experience is sufficient to warrant comparison in only a subset of critical areas. Multiple development routes must be pursued.
2	80% Development Risk—Requires new development where similarity to existing experience can be defined only in the broadest sense. Multiple development routes must be pursued.
1	100% Development Risk—Requires new development outside of any existing experience base. No viable approaches exist that can be pursued with any degree of confidence. Basic research in key areas needed before feasible approaches can be defined.

DESIGN AND ANALYSIS

Do the necessary data bases exist and if not, what level of development is required to produce them?

Do the necessary design methods exist and if not, what level of development is required to produce them?

Do the necessary design tools exist and if not, what level of development is required to produce them?

Do the necessary analytical methods exist and if not, what level of development is required to produce them?

WBS	Element Name	Assessed Area	TRL Score			MRL Score		
			Grn	Yel	Red	Grn	Yel	Red
136905.08.05.03	Structure & Thermal	Primary structure	5	6	7	5		
		Secondary structure	5	6	7	5	6	
		Flight Separation System	5	6, 7	8	5	7	
		TPS	5	6	8	6	7	
136905.08.05.04	Main Propulsion System	Ducts & Lines	5	6	7	5	6	
		Valves & Actuators	5	6, 7	7	5	6	
136905.08.05.05	US Reaction Control System		5	6, 7	8	5	7	
136905.08.05.06	1st Stage Reaction Control System		5	6	8	5	7	
136905.08.05.07	Thrust Vector Control	Actuators	5	6	7	5	6	
		Hydraulic Power Systems	4	5, 6	7	5	6	
136905.08.05.08	Avionics	C&DH	4	5, 6	7	5	6	
		GNC Hardware	5	6	7	6		
		Electrical Power	6	7		6	7	
		Sensors	6	7		6	7	

Figure 12.   AD2 Example.

Do the necessary analysis tools exist and if not, what level of development is required to produce them?

Do the appropriate models with sufficient accuracy exist and if not, what level of development is required to produce them?

Do the available personnel have the appropriate skills and if not, what level of development is required to acquire them?

Has the design been optimized for manufacturability and if not, what level of development is required to optimize it?

Has the design been optimized for testability and if not, what level of development is required to optimize it?

Manufacturing

Do the necessary materials exist and if not, what level of development is required to produce them?

Do the necessary manufacturing facilities exist and if not, what level of development is required to produce them?

Do the necessary manufacturing machines exist and if not, what level of development is required to produce them?

Does the necessary manufacturing tooling exist and if not, what level of development is required to produce it?

Does the necessary metrology exist and if not, what level of development is required to produce it?

Does the necessary manufacturing software exist and if not, what level of development is required to produce it?

Do the available personnel have the appropriate skills and if not, what level of development is required to acquire them?

Has the design been optimized for manufacturability and if not, what level of development is required to optimize it?

Has the manufacturing process flow been optimized and if not, what level of development is required to optimize it?

Has the manufacturing process variability been minimized and if not, what level of development is required to optimize it?

Has the design been optimized for reproducibility and if not, what level of development is required to optimize it?

Has the design been optimized for assembly & alignment and if not, what level of development is required to optimize it?

Has the design been optimized for integration at the component, subsystem and system level and if not, what level of development is required to optimize it?

Are breadboards required and if so what level of development is required to produce them?

Are brassboards required and if so what level of development is required to produce them?

Are subscale models required and if so what level of development is required to produce them?

Are engineering models required and if so what level of development is required to produce them?

Are prototypes required and if so what level of development is required to produce them?

Are breadboards, brassboards, engineering models and prototypes at the appropriate scale and fidelity for what they are to demonstrate, and if not what level of development is required to modify them accordingly?

Are Qualification models required and if so what level of development is required to produce them?

OPERATIONS

Has the design been optimized for maintainability and servicing and if not, what level of development is required to optimize it?

Has the design been optimized for minimum life cycle cost and if not, what level of development is required to optimize it?

Has the design been optimized for minimum annual recurring/operational cost and if not, what level of development is required to optimize it?

Has the design been optimized for reliability and if not, what level of development is required to optimize it?

Has the design been optimized for availability {ratio of operating time (reliability) to downtime (maintainability/ supportability)} and if not, what level of development is required to optimize it?

Do the necessary ground systems facilities & infrastructure exist and if not, what level of development is required to produce them?

Does the necessary ground systems equipment exist and if not, what level of development is required to produce it?

Does the necessary ground systems software exist and if not, what level of development is required to produce it?

Do the available personnel have the appropriate skills and if not, what level of development is required to acquire them?

TEST & EVALUATION

Do the necessary test facilities exist and if not, what level of development is required to produce them?

Does the necessary test equipment exist and if not, what level of development is required to produce them?

Does the necessary test tooling exist and if not, what level of development is required to produce it?

Do the necessary test measurement systems exist and if not, what level of development is required to produce them?

Does the necessary software exist and if not, what level of development is required to produce it?

Do the available personnel have the appropriate skills and if not, what level of development is required to acquire them?

Has the design been optimized for testability and if not, what level of development is required to optimize it?

Are breadboards required to be tested and if so what level of development is required to test them?

Are brassboards required to be tested and if so what level of development is required to test them?

Are subscale models required to be tested and if so what level of development is required to test them?

Are engineering models required to be tested and if so what level of development is required to test them?

Are prototypes required to be tested and if so what level of development is required to test them?

Are Qualification models required to be tested and if so what level of development is required to test them?

Each element is evaluated according to the scale in Table 1. The resulting scores are tabulated at the far right of the table to indicate how many categories are at what level. The penultimate column provides a numerical assessment of the overall difficulty, and the final column displays a color-coded assessment corresponding to a level of concern. In scoring the table, the degree of difficulty is never higher than the lowest value in the row. However, the degree of difficulty may be increased further by the presence of multiple categories having low values. An algorithm for calculating AD^2 scores is given in Attachment C. It should be carefully noted that the value in the AD^2 process is in the generation of the data, it cannot be easily reduced to a single number and consequently such reductions are intended to be used as general indications of difficulty.

Again it is important to remember that the questions should be tailored to the project and the level of detail should be appropriate to the phase of the program. Many of the questions outlined above can be expanded upon, e.g. in the area of design:

Design Life: related to wear-out and replacement
Identification of key critical functions/failures and design solutions identified through testing and demonstration
Ergonomics: human limitations and safety considerations
Reduction in complexity
Duplication to provide fault tolerance
Is derating criteria established to limit stress on components?
Are Modularization concepts applied?
Commonality with other systems
Reconfigurability
Feedback of failure information (lessons learned, reliability predictions, failure history, etc.) to influence design
Maintenance philosophy
Storage Requirements
Transportation Requirements
Design for parts orientation and handling
Design for ease of assembly

Or in manufacturing:

Integrated Design/manufacturing tools?
Materials? Are they compatible with manufacturing technology selected?
Non-destructive evaluation? Other inspections?
Workforce (with right skills) availability?
Can building-block approach be followed?

Or in operations:

Maximized Launch Readiness Probability?

o Related to Availability but covers the period of time from start of ground processing (e.g. start of launch vehicle stacking) to start of launch countdown (e.g. "LCC Call to Station").

o For example: Launch Readiness of 85% or greater—Ability to be ready for launch countdown 85% of the time or better on the first scheduled launch attempt (excluding weather and ground systems, mission systems, and payload/spacecraft anomalies)

o Variables that influence the launch readiness probability:

- Availability
- MTBF
- MTTR
- Ability to meet H/W delivery dates

Successful determination of AD^2 in this process requires that personnel with unique skills be added to the team who performed the initial TRL assessment. The TRL assessment required knowledge of what had and had not been done to date. It is imperative that these individuals <u>know</u> what has happened in the field, but they are not necessarily experts. In order for the AD^2 assessment to have any validity, experts must perform the assessment. It is for all intents and purposes a predictive undertaking, and since crystal balls are in short supply, the only rational approach is to employ individuals that have the background and experience necessary to provide realistic assessments and projections. It is rare that any one individual (or even two individuals) will possess the requisite knowledge to cover all areas, and so expertise must be drawn from throughout the community.

Occasionally the required expertise does not exist at all, meaning that diverse groups will have to be put together to form a collective opinion. An example of this might involve a laser in which the laser researchers have no experience in space qualification and the personnel with space qualification experience have no experience with lasers. The only solution to this is to get the two groups together for an extended period of time, long enough that they can each understand the issues as a collective group. Establishing the proper assessment group is one of the most challenging activities that a program/project must face and one most critical to success.

An automated version of an AD^2 calculator is under development and is shown in Figure 13. A copy is included with the companion software that accompanies this book.

Schedule	Cost	AD2 Level	Questions
			Design and Analysis
			Do the necessary data bases exist and if not, what level of development is required to produce them?
			Do the necessary design methods exist and if not, what level of development is required to produce them?
			Do the necessary design tools exist and if not, what level of development is required to produce them?
			Do the necessary analytical methods exist and if not, what level of development is required to produce them?
			Do the appropriate models with sufficient accuracy exist and if not, what level of development is required to produce them?
			Do the available personnel have the appropriate skills and if not, what level of development is required to acquire them?
			Has the design been optimized for manufacturability and if not, what level of development is required to optimize it?

Figure 13.   Automated AD2 Calculator.

It is expected that the assessment both of the TRL and the AD^2 will be accomplished from within the program/project with program/project personnel augmented as necessary when critical skills are absent. In order to validate this assessment it will be important to establishing an Independent Advisory Panel (IAP) to periodically review the assessment process. There are many pitfalls in maturing technology and even the most experienced program/project manger with the best team in the world will benefit from the advice of an IAP. An outside group of experts can provide insight into the direction and progress in a positive, constructive and "non-threatening" manner. This insight will help keep the program on track and maximize the probability of success. The advice from this panel will be of extreme importance throughout the duration of the program; however, its greatest impact will be in the evaluation of the initial set of goals/requirements and the results of the Assessment Teams that will identify the key technologies, establish their readiness levels and determine the degree of difficulty involved in advancing them to the requisite readiness levels. These results form the basis for success of the entire program/project. They are used to build roadmaps, establish priorities, estimate costs, and create development paths, milestones and schedules. It is incredibly important that these assessments are part of the base-lining process. Once the initial reviews have been conducted and the results assimilated, the IAP should be called upon to periodically assess results at critical points throughout the life of the program/project through PDR.

The makeup of the advisory panel is very important, and considerable time and effort should be expended in making sure that the proper mix of expertise is included. The panel should be comprised of very senior people who have no vested interest in the program/project. The panel should have as broad a range of experience as possible and contain genuine experts in all the critical technology areas. As a minimum the panel should include a senior manager with experience in managing technology programs/projects, a senior manager with experience in managing flight hardware programs/projects, a senior technologist, a senior systems engineer and a senior operations engineer. Depending upon the program/project, the panel should also include a senior member from the Department of Defense or any other agency that has similar activity underway. It should also include discipline experts drawn from the academic and industrial communities.

The IAP lead must be highly respected in the community with a broad range of experience and, most importantly, the ability to devote the necessary time to the program, particularly in the early stages. The role of the lead is to provide straight, unvarnished, objective advice relative to the program/project. Consequently, it is important that you choose a lead that is capable of such interaction.

ESTABLISHING MILESTONES AND
TECHNICAL PERFORMANCE MEASURES
RELATIVE TO TECHNOLOGY MATURATION

The purpose of establishing milestones is to enable progress to be tracked, anticipate problems as far in advance as possible and to make timely decisions relative to alternative paths and fall-back positions. They also allow external entities to track progress, and therein is the problem. It is extremely important in the technology maturation process that these milestones have quantifiable metrics associated with them. These time-phased metrics are known as Technical Performance Measures (TPMs). There is a natural tendency to call out milestones and define metrics that are easily met in order to "protect" the activity from unwanted outside influence. Unfortunately, there is a severe downside to this approach. Milestones and metrics that do not provide any insight to outside micro-managers also do not provide any insight to the program/project manager and that is not good for the program/project. The only true alternative is to take the time to establish metrics and milestones that will be of value in assuring that the technological maturation required by the program/project occurs in a timely and effective manner and deal with unwanted advice when it comes.

Establishing good, quantifiable metrics/milestones is an art that requires in-depth knowledge of what is being undertaken, its current state and where it is supposed to be going. In other words, the information provided by the TRL and AD^2 assessments. It is extremely important to emphasize the word <u>quantifiable</u>! *If you can't measure it—you can't make it!* The heart of the issue of progress measurement is testing, the most important aspects of measuring progress. Once it accepted that "testing" is the underpinning of the way to measure progress, the definition of appropriate metrics becomes somewhat more tractable. One of the most difficult areas to track progress today is in the area of software development, whether it is in a flight program or a technology program. Lines of code do not tell you what "functions" are being developed. Neither do the number of "builds," since certain functions are often deferred to the next build. What is quantifiable is the function a given set of software can perform. In other words what would be the result if it were *tested*? Difficult problems must be broken down into manageable sized "chunks" that have (1) reasonably well understood technical goals, (2) short enough development times such that success or failure can have the appropriate impact on the overall effort, and (3) measurable output that demonstrates appropriate incremental progress. As was the case with cost, schedules and milestones, the data generated during the AD^2 assessment provides important insight into establishing the appropriate metrics required.

OUTPUTS/PRODUCTS/EXIT CRITERIA

Technology Development Plan

The Technology development plan identifies key technological advances and describes the steps necessary to bring them to a level of maturity (TRL) that will permit them to be successfully integrated into a program/project. It provides the overall direction of the effort. The plan is developed after the completion of the TMA and AD^2 assessments. These assessments also provide critical data for program costing, scheduling and implementation planning. The Technology Assessment process identifies through the TMA the maturity of technologies required to be incorporated into the program, and the AD^2 assessment establishes the requirements for maturing those technologies. Once these difficult processes have been successfully completed, generation of the plan is simply documentation of the results—a hierarchical collection of maps that starts at the highest system level and follows the breakdown into subsystems and components as established in the TRL assessment. The AD^2 assessment is used to determine where parallel approaches should be put in place. Again, in order to maximize the probability of success it will be highly desirable to have multiple approaches to highly difficult activities (to the extent possible within cost constraints). The AD^2 assessment also provides considerable insight into what breadboards, engineering models and or prototypes will be needed and what type of testing and test facilities will be needed. The AD^2 assessment plays a principal role in establishing program costs and schedules, and it is impossible to overemphasize the importance of having an accurate assessment as early as possible. A preliminary version of the Technology Development Plan is required for transition from Pre-phase A to Phase A. The final version of the plan is required for transition from Phase A to Phase B.

TECHNOLOGY ASSESSMENT REPORT

The Technology Assessment Report is prepared at the end of Phase B and presented at the PDR. It serves to document that all systems, subsystems and components have been demonstrated through test and analysis to be at a maturity level at or above TRL6. The TAR is required for transition from Phase A to Phase C.

APPENDIX D BIBLIOGRAPHY

Schinasi, Katherine, V., Sullivan, Michael, "Findings and Recommendations on Incorporating New Technology into Acquisition Programs," Technology Readiness and Development Seminar, *Space System Engineering and Acquisition Excellence Forum*, The Aerospace Corporation, April 28, 2005.

"Better Management of Technology Development Can Improve Weapon System Outcomes," GAO Report, GAO/NSIAD-99-162, July 1999.

"Using a Knowledge-Based Approach to Improve Weapon Acquisition," GAO Report, GAO-04-386SP. January 2004.

"Capturing Design and Manufacturing Knowledge Early Improves Acquisition Outcomes," GAO Report, GAO-02-701, July 2002.

Wheaton, Marilee, Valerdi, Ricardo, "EIA/ANSI 632 as a Standardized WBS for COSYSMO," 2005 NASA Cost Analysis Symposium, April 13, 2005.

Sadin, Stanley T.; Povinelli, Frederick P.; Rosen, Robert, "NASA technology push towards future space mission systems," Space and Humanity Conference Bangalore, India, Selected Proceedings of the 39th International Astronautical Federation Congress, Acta Astronautica, pp 73–77, V 20, 1989

Mankins, John C. "Technology Readiness Levels" a White Paper, April 6, 1995.

Nolte, William, "Technology Readiness Level Calculator, "Technology Readiness and Development Seminar, *Space System Engineering and Acquisition Excellence Forum*, The Aerospace Corporation, April 28, 2005.

TRL Calculator is available at the Defense Acquisition University Website at the following URL: https://acc.dau.mil/communitybrowser.aspx?id=25811

Manufacturing Readiness Level description is found at the Defense Acquisition University Website at the following URL: https://acc.dau.mil/CommunityBrowser.aspx?id=25811>

Mankins, John C., "Research & Development Degree of Difficulty (RD^3)" A White Paper, March 10, 1998.

Sauser, Brian J., "Determining system Interoperability using an Integration Readiness Level," Proceedings, NDIA Conference Proceedings.

Sauser, Brian, Ramirez-Marquez, Jose, Verma, Dinesh, Gove, Ryan, "From TRL to SRL: The Concept of Systems Readiness Levels, Paper #126, Conference on Systems Engineering Proceedings.

For information on automated AD^2 contact james.w.bilbro@nasa.gov

Additional Material of Interest

De Meyer, Arnould, Loch, Christoph H., and Pich Michael T., "Managing Project Uncertainty: From Variation to Chaos," MIT Sloan Management Review, pp. 60–67, Winter 2002.

ATTACHMENT A:
TECHNOLOGY DEVELOPMENT TERMINOLOGY

Proof of Concept: (TRL 3)—Analytical and experimental demonstration of hardware/software concepts that may or may not be incorporated into subsequent development and/or operational units.

Breadboard: (TRL 4)—A low fidelity unit that demonstrates function only, without respect to form or fit in the case of hardware, or platform in the case of software. It often uses commercial and/or ad hoc components and is not intended to provide definitive information regarding operational performance.

Developmental Model/Developmental Test Model: (TRL 4)—Any of a series of units built to evaluate various aspects of form, fit, function or any combination thereof. In general these units may have some high fidelity aspects but overall will be in the breadboard category.

Brassboard: (TRL 5–TRL 6)—A mid-fidelity functional unit that typically tries to make use of as much operational hardware/software as possible and begins to address scaling issues associated with the operational system. It does not have the engineering pedigree in all aspects, but is structured to be able to operate in simulated operational environments in order to assess performance of critical functions.

Mass Model: (TRL 5)—Nonfunctional hardware that demonstrates form and/or fit for use in interface testing, handling, and modal anchoring.

Subscale model: (TRL 5–TRL7)—Hardware demonstrated in subscale to reduce cost and address critical aspects of the final system. If done at a scale that is adequate to address final system performance issue it may become the prototype.

Proof Model: (TRL 6)—Hardware built for functional validation up to the breaking point, usually associated with fluid system over pressure, vibration, force loads, environmental extremes, and other mechanical stresses.

Prototype Unit: (TRL 6–TRL 7)—The prototype unit demonstrates form (shape and interfaces), fit (must be at a scale to adequately address critical full size issues), and function (full performance capability) of the final hardware. It can be considered as the first Engineering Model. It does not have the engineering pedigree or data to support its use in environments outside of a controlled laboratory environment – except for instances where a specific environment is required to enable the functional operation including in-space. It is to the maximum extent possible identical to flight hardware/software and is built to test the manufacturing and testing processes at a scale that is appropriate to address critical full scale issues.

Engineering Model: (TRL 6–TRL 8)—A full scale high-fidelity unit that demonstrates critical aspects of the engineering processes involved in the development of the operational unit. It demonstrates function, form, fit or any combination thereof at a scale that is deemed to be representative of the final product operating in its operational environment. Engineering test units are intended to closely resemble the final product (hardware/software) to the maximum extent possible and are built and tested so as to establish confidence that the design will function in the expected environments. In some cases, the engineering unit will become the protoflight or final product, assuming proper traceability has been exercised over the components and hardware handling.

Flight Qualification Unit: (TRL 8)—Flight hardware that is tested to the levels that demonstrate the desired margins, particularly for exposing fatigue stress., typically 20-30%. Sometimes this means testing to failure. This unit is never flown. Key overtest levels are usually +6db above maximum expected for 3 minutes in all axes for shock, acoustic, and vibration; thermal vacuum 10C beyond acceptance for 6 cycles, and 1.25 times static load for unmanned flight.

Protoflight Unit: (TRL 8–TRL 9)—Hardware built for the flight mission that includes the lessons learned from the Engineering Model but where no Qualification model was built to reduce cost. It is however tested to enhanced environmental acceptance levels. It becomes the mission flight article. A higher risk tolerance is accepted as a tradeoff. Key protoflight overtest levels are usually +3db for shock, vibration, and acoustic; 5C beyond acceptance levels for thermal vacuum tests.

Flight Qualified Unit: (TRL 8–TRL 9)—Actual flight hardware/software that has been through acceptance testing. Acceptance test levels are designed to demonstrate flight-worthiness, to screen for infant failures without degrading performance. The levels are typically less than anticipated levels.

Flight Proven: (TRL 9)—Hardware/software that is identical to hardware/software that has been successfully operated in a space mission.

Environmental Definitions

Laboratory Environment—An environment that does not address in any manner the environment to be encountered by the system, subsystem or component (hardware or software) during its intended operation. Tests in a laboratory environment are solely for the purpose of demonstrating the underlying principles of technical performance (functions) without respect to the impact of environment.

Relevant Environment: Not all systems, subsystems and/or components need to be operated in the operational environment in order to satisfactorily address performance margin requirements. Consequently, the relevant environment is the specific subset of the operational environment that is required to demonstrate critical "at risk" aspects of the final product performance in an operational environment.

Operational Environment—The environment in which the final product will be operated. In the case of spaceflight hardware/software it is space. In the case of ground based or airborne systems that are not directed toward space flight it will be the environments defined by the scope of operations. For software, the environment will be defined by the operational platform and software operating system.

Additional Definitions

Mission Configuration—The final architecture/system design of the product that will be used in the operational environment. If the product is a subsystem/component then it is embedded in the actual system in the actual configuration used in operation.

Validation—Demonstration by test that a device meets its functional and environmental requirements (i.e., did I build the thing right?).

Verification—Determination that a device was built in accordance with the totality of its prescribed requirements by any appropriate method. Commonly uses a verification matrix of requirement and method of verification (i.e., did I build the right thing?).

Part—Single piece or joined pieces impaired or destroyed if disassembled –e.g., a resistor.

Subassembly or **component**—Two or more parts capable of disassembly or replacement—e.g., populated printed circuit board.

Assembly or Unit – a complete and separate lowest level functional item —e.g., a valve.

Subsystem—Assembly of functionally related and interconnected units— e.g., electrical power subsystem.

System—The composite equipment, methods, and facilities to perform and operational role.

Segment—The constellation of systems, segments, software, ground support, and other attributes required for an integrated constellation of systems.

ATTACHMENT C: AD² WEIGHTING ALGORITHM

Note, reducing the AD^2 process to a single number should be viewed very cautiously. The important information is gained in the process, and cannot be adequately described in a single number.

Table 1. Attachment B: TRL Descriptions

Technology Readiness Level (TRL)	Definition	Hardware Description	Software Description	Exit Criteria
1	Basic principles observed and reported	Scientific knowledge generated underpinning hardware technology concepts/applications.	Scientific knowledge generated underpinning basic properties of software architecture and mathematical formulation	Peer reviewed publication of research underlying the proposed concept/application
2	Technology concept and/or application formulated	Invention begins, practical application is identified but is speculative, no experimental proof or detailed analysis is available to support the conjecture.	Invention begins. Practical application exists but is speculative. No experimental proof or detailed analysis is available to support the conjecture. Underlying algorithms are created and documented.	Documented description of the application / concept that addresses feasibility and benefit.
3	Analytical and/or experimental critical functions or characteristic proof-of-concept	Analytical studies place the technology in an appropriate context and laboratory demonstrations, modeling and simulation validate analytical predictions.	Development of limited functionality to validate critical properties and predictions using non-integrated software components	Documented analytical/ experimental results validating predictions of key parameters
4	Component or breadboard validation in laboratory	A low fidelity system/component breadboard is built and operated to demonstrate basic functionality and critical test environments and associated performance predictions are defined relative to the final operating environment.	Key, functionally critical, software components are integrated, and functionally validated, to establish interoperability and begin architecture development. Relevant environments defined and performance in this environment predicted.	Documented test performance demonstrating agreement with analytical predictions. Documented definition of relevant environment.

Table continues on next page.

5	Component or breadboard validation in a relevant environment	A mid-level fidelity system/ component brassboard is built and operated to demonstrate overall performance in a simulated operational environment with realistic support elements that demonstrates overall performance in critical areas. Performance predictions are made for subsequent development phases.	End to end software elements implemented and interfaced with existing systems conforming to target environment, including the target software environment. End to end software system tested in relevant environment meets predicted performance. Operational environment performance predicted.	Documented test performance demonstrating agreement with analytical predictions. Documented definition of scaling requirements.
6	System/subsystem model or prototype demonstration in a relevant environment	A high fidelity system/ component prototype that adequately addresses all critical scaling issues is built and operated in a relevant environment to demonstrate operations under critical environmental conditions.	Prototype software partially integrated with existing hardware/ software systems and demonstrated on full-scale realistic problems.	Documented test performance demonstrating agreement with analytical predictions.
7	System prototype demonstration in space	A high fidelity engineering unit that adequately addresses all critical scaling issues is built and operated in a relevant environment to demonstrate performance in the actual operational environment and platform (ground, airborne, or space).	Prototype software is fully integrated with operational hardware/software systems demonstrating operational feasibility.	Documented test performance demonstrating agreement with analytical predictions.
8	Actual system completed and flight qualified through test and demonstration	The final product in its final configuration is successfully demonstrated through test and analysis for its intended operational environment and platform (ground, airborne or space).	The final product in its final configuration is successfully [demonstrated] through test and analysis for its intended operational environment and platform (ground, airborne or space).	Documented test performance verifying analytical predictions.
9	Actual system flight proven through successful mission operations	The final product is successfully operated in an actual mission.	The final product is successfully operated in an actual mission.	Documented mission operational results.

If any category of any system, subsystem or component is at an AD^2 Level 4 or below then the system, subsystem and component are all at that level.

If all of the categories are above AD^2 Level 4, then:

$$\text{Composite } AD^2 = \frac{[\sum \text{Category Level X Probability of Occurrence}]}{\# \text{Categories}}$$

The color code associated with each row provides a visual cue relative to the degree of difficulty:

Red: High Risk 1–4
Yellow: Moderate Risk 5–6
Green: Low Risk 7–9
White: Not considered

© 2006 M. Engling

Appendix E

Before you install the TRL Calculator from the accompanying software, read the file named "Version Selection Guide." This document will help you decide which version of the calculator is right for you.

TRL Calculator Version 2.2, found in the "Version 2.2" folder of the software that accompanies this book, is the latest version of the AFRL TRL calculator. This version has been cleared for public release. It works well with Microsoft Excel 2003. Many of the controls activate macros that perform the actual function, such as hiding questions that aren't used when you tailor the calculator to your technology program. I have not tried it with Vista.

If you're using Excel 2000 or earlier, you'll probably get some run-time errors when you use some of the features. That's because some of the macros in version 2.2 use commands that are not supported in earlier versions of Excel, so clicking on the controls gives annoying run-time errors. Disabling these features in TRL Calculator Version 2.21 eliminated the run-time errors. If you are running MS Excel 2000 or earlier, you'll need this version, found in the folder "Version 2.21." While the display looks slightly different, because unused rows aren't hidden, the functionality is unchanged.

The NASA version is in a separate folder. It is a variation of the TRL Calculator that has been tailored to eliminate questions peculiar to DoD. The remaining questions were reworded to match NASA nomenclature, and then placed in order of increasing importance. The NASA variant also groups hardware, software, and manufacturing questions together at each TRL. Use this version if you're working for or with NASA unless directed otherwise. The "NASA Version" folder also contains a copy of the automated tool for performing an advancement degree of difficulty (AD2) assessment, since the AD2 Calculator has been linked to the NASA TRL Calculator. The software includes notional example data, not taken from any real technology development program or project, to illustrate the tool and some of its features. A word of caution: don't alter the example data, and don't alter any saved data you may generate. Let the spreadsheet's macros and links take care of the data for you to avoid corrupting the data files.

When you install any version of the calculator, you need to make sure that Excel is set to Medium security. From the top line drop-down menu, select "Tools," then "Macros" and "Security." Pick the "Medium" radio button. When you open a calculator spreadsheet containing macros, a dialog box will pop up asking whether or not you wish to allow macros to run. The default selection is "Disable." You'll need to change it to "Enable" at the time you load the spreadsheet, or some of the features won't work.

Install all of the files into a single directory or folder, and the links should work. Some users have said that after unzipping, they needed to activate each file manually once before the links worked correctly, so you might try that if the links don't appear to work at first.

Start by selecting the file called "Documentation 2.2.xls" to view background information on TRLs, or plunge right in by selecting the file "TRL Calc Ver 2.2.xls." Follow the links from there. A more complete set of instructions and a full description of the calculator can be found in the "Release Notes" included in the documentation. The release notes can be reached from either spreadsheet.

I certify that the macros contained in the software files that come with this book are virus free.

© 2006 M. Engling

INDEX

189

Supplemental information, 68-75
Supportability, 5, 10, 12
Supporting information, 49, 50-55, 56-57
Sustainability, 12
Sustainment/supportability readiness levels, 87-88
System development and demonstration phase, 26-27

T
Taleb, Nassim Nicholas, 94
Technical assessment, xvii, xviii, 152, 154-156, 156-159
Technical performance measure (TPM) risk index, 106, 108
Technical performance risk index (TPRI), 108-109
Technologist, 2
Technology, 5-6
Technology demonstration environment, 56
Technology developer, 3, 57
Technology development, 10, 11, 152
Technology development phase, 26
Technology development plan (NASA), 155, 178
Technology development strategy (TDS), 26
Technology growth, 7
Technology immaturity, xv, xvii, 3
Technology, kinds of, 6
Technology life cycle, xxii, 2, 17, 19, 20-23, 24, 28, 32, 63
Technology maturity, xi, xiii, xv, xvii, xxii, 1, 2-5, 7, 9-17, 25, 26, 28, 45, 46, 57-58, 59, 61, 62-63, 66,45, 46, 57-58, 59, 61, 62-63, 66, 78, 83, 93, 119, 120, 153, 155
Technology maturity assessment (TMA),113, 114, 118, 154-155, 178
Technology maturity dimensions, 10-17
Technology, non-material, 6
Technology readiness assessment, xvii, 47, 50, 56, 57, 64, 116, 117, 153

Technology readiness assessment report (NASA), 156, 178
Technology readiness level (TRL), xi, xiii, xv, xvi, xvii, 1, 11, 45-59, 63, 64, 67, 78, 79, 80, 84, 87, 89, 90, 91, 99, 106, 109, 116, 117, 155, 156, 158-163, 164, 167-168, 176, 177, 178, 188
  TRL assessment, 163, 168, 176
  TRL table, 48, 50-55, 90, 160, 183-184
Technology user, 3, 5
Test and evaluation, 81, 110, 111, 172
Trade studies, 153
Traditionalists, 37
Transition, xv, 14, 21, 57-58, 81, 89, 135, 155, 178
TRL calculator, xi, xvii, 56, 57, 66, 80, 117, 164-167, 188-189
TRL definitions, 48, 49, 56
  TRL 1, 50, 68
  TRL 2, 50-51, 68-69
  TRL 3, 90, 69, 180
  TRL 4, 51-52, 70, 180
  TRL 5, 52-53, 71, 180
  TRL 6, 53, 72, 180
  TRL 7, 53-54, 73, 180
  TRL 8, 54, 74, 180, 181
  TRL 9, 55, 75, 181
TRL limitations, 58-59
TRL uses, 57-58
Turner, Richard, 9, 63, 76

U
Uncertainty, xvii-xviii
Uncertainty profile, xviii

V
Variation, xviii, 93-94, 95, 96-98, 98-99, 100-101, 101-102, 102-103, 131-135, 136, 137, 142
Variation estimation, 98-99, 131-148
  Assumptions, 101, 136
    Confidence interval, 136-137
    Gaussian, 101-102, 136, 137-138
    Range = 4 sigma, 141-142
  Estimation methods, 137-147